Nonlinear Systems - Practical Applications and Contemporary Challenges

Edited by Ricardo López-Ruiz

Published in London, United Kingdom

Nonlinear Systems - Practical Applications and Contemporary Challenges
http://dx.doi.org/10.5772/intechopen.1007995
Edited by Ricardo López-Ruiz

Contributors
Alain-Brice Moubissi, Andrei V. Metrikine, Apostolos Tsouvalas, Boka Kumsa, Byron J. Idrovo-Aguirre, Héctor Acuña, J. Omana, Jyotrimoy Rana, Manzoor Ahmad Lone, Maïk Delon Mboumba, Owen Casha, Pablo Easton, Panagiota Atzampou, Pedro Eduardo Alvarado Rubio, Peter C. Meijers, Ricardo Lino Mansilla Corona, Ricardo López-Ruiz, Sarita Kumari, Shaima Qureshi, Tarekegn Dinku Gelu, Thierry Blanchard Ekogo

First published in London, United Kingdom, 2026 by IntechOpen
IntechOpen is the global imprint of INTECHOPEN LIMITED, registered in England and Wales, registration number: 11086078, 167-169 Great Portland Street, London, W1W 5PF, United Kingdom

For EU product safety concerns: IN TECH d.o.o., Prolaz Marije Krucifikse Kozulić 3, 51000 Rijeka, Croatia, info@intechopen.com or visit our website at intechopen.com.

British Library Cataloguing-in-Publication Data
A catalogue record for this book is available from the British Library

Nonlinear Systems - Practical Applications and Contemporary Challenges
Edited by Ricardo López-Ruiz
p. cm.
Print ISBN 978-1-83635-659-2
Online ISBN 978-1-83635-658-5
eBook (PDF) ISBN 978-1-83635-660-8

If disposing of this product, please recycle the paper responsibly.

Meet the editor

Ricardo López-Ruiz, MS, Ph.D., is an associate professor in the Department of Computer Science and Systems Engineering, Faculty of Science, University of Zaragoza, Spain. He is also an associate researcher in Complex Systems at the University of Zaragoza's School of Mathematics. Previously, he worked as a lecturer at the University of Navarra, the Public University of Navarra, and UNED of Calatayud. He completed his postdoc with Prof. Yves Pomeau at the École Normale Supérieure of Paris and with Prof. Gabriel Mindlin at the University of Buenos Aires. His areas of interest include statistical complexity and nonlinear models, chaotic maps and applications, multi-agent systems, econophysics, big data, and artificial intelligence techniques.

Contents

Preface

Nonlinear systems present a set of challenges, primarily due to their complex and unpredictable behavior. Unlike linear systems, where inputs and outputs maintain a proportional relationship, nonlinear systems can exhibit phenomena such as chaos, bifurcations, and multistability. This makes predicting system behavior challenging, as minor changes in inputs or initial conditions can yield vastly different outcomes.

Another significant challenge is controlling and stabilizing nonlinear systems, especially in practical applications such as robotics, energy systems, and climate modeling. Over three sections, this book aims to delve into this type of system from theoretical approaches to practical applications. A wide range of topics, including analysis of stability and bifurcations, the control of the dynamics, and the prediction of the behavior of these systems, is covered in different biological, physical, and theoretical situations, as is done in the introductory chapter for the case of nonlinear oscillators of Liénard type and different models built with these elements.

The second section of the book presents different studies related to the behavior of the illness in patients analyzed from the point of view of embedding dimension and permutation (Chapter 2 by Dr. Alvarado-Rubio et al.), the mathematical analysis of a food chain model to illustrate some ecological phenomenon (Chapter 3 by Dr. Kumari and Omana), the possibility of controlling the formation and dynamics of soliton molecules in binary dipolar Bose–Einstein condensates with spin–orbit coupling (Chapter 4 by Dr. Delon Mboumba et al.), and the contactless magnetic control strategy designed to simultaneously manage the translational and torsional motions of a suspended load (Chapter 5 by Dr. Atzampou et al.).

The third section of the book addresses the potential of Caputo-Fabrizio derivatives to enhance the understanding of tumor growth dynamics (Chapter 6 by Dr. Dinku Gelu et al.), the application of describing functions in predicting the behavior of nonlinear oscillators (Chapter 7 by Dr. Casha), and the use of a nonlinear 2D logistic chaotic map is discussed efficiently and effectively to encrypt digital images (Chapter 8 by Dr. Manzoor Ahmad Lone and Shaima Qureshi) and the demonstration that ensuring the continuous development of infrastructure investment plans helps mitigate inefficiencies in capital formation caused by bureaucratic constraints (Chapter 9 by Dr. Idrovo-Aguirre et al.).

As the editor of this book, I would like to thank all the contributors who have helped build this book through their work and assessments. Also, I am grateful to the IntechOpen Editorial Staff for their invitation to serve as editor for the eighth time, and for the particular and efficient help of Ms. Marica Novakovic, the Publishing Process Manager, in converting this new book. Finally, at a time when wars continue to devastate parts of the world, I want to dedicate this book to all the kind and friendly people I have met in my life, who made my time with them a marvelous experience. Of course, all my family, friends, and past advisors are not forgotten in this dedicatory final paragraph.

Ricardo López-Ruiz
University of Zaragoza,
Spain

Section 1

Introduction

Chapter 1

Introductory Chapter: Analytical Approaches and Contemporary Challenges in Nonlinear Systems

Ricardo López-Ruiz

1. Introduction

Nonlinear oscillators of Liénard type form a fundamental class of dynamical systems describing self-sustained oscillations in physics, biology, engineering, and electronics. A classical example of Liénard systems is the van der Pol (vdP) oscillator,

$$\ddot{x} + \varepsilon\left(x^2 - 1\right)\dot{x} + x = 0 \tag{1}$$

which is the prototypical example exhibiting a unique, stable limit cycle whose shape ranges from near-harmonic ($\varepsilon \to 0$) to relaxation-type ($\varepsilon \to \infty$) oscillations. Despite extensive study, obtaining analytical approximations valid across the full nonlinear regime remains a challenging problem.

Here, we address this challenge by reviewing the application of the *Homotopy Analysis Method* (HAM) [1] that provides a general framework for arbitrary Liénard systems and allows to compute accurately amplitudes, frequencies, and limit-cycle shapes. In the case of the vdP, it produces closed-form amplitude approximations valid for all $\varepsilon > 0$. This chapter synthesizes these methods and results, and highlights remaining challenges in nonlinear systems theory. Also, different nonlinear models built from the vdP oscillator are recalled in this short review.

2. Liénard systems and limit cycles

A Liénard equation has the general form

$$\ddot{x} + \varepsilon f(x)\dot{x} + x = 0 \tag{2}$$

with $f(x)$ typically an even polynomial. The existence and number of limit cycles remain central questions in dynamical systems [2]. Classical results cover low-degree cases for $f(x)$, but for degree 6 or higher there exist counterexamples to the well-known Lins–Melo–Pugh conjecture [3].

A useful transformation eliminates the time variable by setting $v = \dot{u}$ and treating v as a function of u. The resulting equation

$$vv' + \varepsilon f(u)v + u = 0 \tag{3}$$

can be rescaled to a fixed domain. In this form, HAM can be applied directly in the (u, v) phase plane. The initial guess used in the homotopy reduces exactly to the circular orbit as $\varepsilon \to 0$ and to the relaxation oscillation structure as $\varepsilon \to \infty$ giving a uniformly valid starting point.

2.1 The Homotopy analysis method for Liénard equations

HAM introduces an embedding parameter $p \in [0, 1]$ to continuously deform an initial guess into the true solution [4]. Unlike perturbation methods, HAM:

- does not require a small parameter,
- provides convergence control *via* an auxiliary parameter h,
- generates linear recursive problems at each order.

2.1.1 Time-domain formulation

A standard approach [5] rescales time as $\tau = \omega t$ and the displacement as $x = a\, u(\tau)$. The solution $u(\tau)$ is expanded in Fourier series, and solvability conditions eliminate secular terms. HAM yields recursive corrections to the amplitude a, frequency ω, and waveform. Classical perturbation results appear naturally when $h = -1$, while other choices of h produce convergent series even for large ε.

2.1.2 Phase-space formulation for the van der pol oscillator

For the vdP oscillator, a distinct HAM formulation [6] is constructed directly in the $(x, \dot{x})$ phase plane. The homotopy generates corrections $y_0(x), y_1(x), y_2(x)$, ... to the limit-cycle profile $y(x)$, as well as corrections to the amplitude a. With a properly chosen $h(\varepsilon)$, the resulting series converge quickly and uniformly.

2.2 Analytical amplitude formulas for the van der pol oscillator

Historically, only perturbation expansions (valid for small ε) and asymptotic expansions (valid for large ε) were available [7, 8]. HAM provides the first globally valid formulas for the amplitude.

2.2.1 First-order approximation

Using $h(\varepsilon) = \frac{b\varepsilon}{c+\varepsilon^2}$ with suitable constants, $b = -1.3$ and $c = 4$, the following closed-form approximation $a_{R1}(\varepsilon)$ for the amplitude of the vdP oscillator is obtained [6]:

$$a_{R1}(\varepsilon) = 2 + \frac{1.737\,\varepsilon^2}{(8\pi + 9\varepsilon)\,(4 + \varepsilon^2)}. \tag{4}$$

This formula matches numerical results with a relative error below 0.1% for all $\varepsilon > 0$.

2.2.2 Second-order refined approximation

Including second-order corrections for the same $h(\varepsilon)$ with suitable constants, $b = -0.561$ and $c = 4$, yields a more accurate expression $a_{R2}(\varepsilon)$ for the approximation of the amplitude of the vdP oscillator [6]:

$$a_{R2}(\varepsilon) = 2 + \frac{0.74958\,\varepsilon^2}{(8\pi + 9\varepsilon)\,(4 + \varepsilon^2)} + \frac{\varepsilon^2(75.3562 + 43.0023\varepsilon + 28.1589\varepsilon^2 + 8.3479\varepsilon^3)}{(8\pi + 9\varepsilon)^2(4 + \varepsilon^2)^2}. \tag{5}$$

This achieves accuracy better than 0.05% across the entire parameter range. Numerical evidence suggests $a_{R2}(\varepsilon)$ may be an upper bound of the true amplitude.

2.3 Reconstruction of limit-cycle shapes

The phase-space HAM formulation also produces accurate approximations of the waveform $y(x) = \dot{x}$ along the limit cycle. Already the second-order approximation reproduces the fast–slow features of relaxation oscillations (e.g., at $\varepsilon > 5$) with excellent agreement with numerical integration [9]. This is notable because waveform reconstruction is generally more difficult than amplitude estimation and even than finding the bifurcation curves of the limit cycles [10].

2.4 Broader implications and contemporary challenges

These works reveal HAM as a powerful analytic tool capable of bridging weakly and strongly nonlinear regimes. Nevertheless, several important challenges remain [11, 12]:

- Classification of multiple limit cycles in general Liénard systems.
- Analytical treatment of high-degree polynomial nonlinearities, where classical conjectures fail.
- Integration of HAM-based reduced models into engineering control and design.
- Development of rigorous convergence theory for HAM, including quantitative criteria for selecting h.

3. Oscillation modes in a model of two coupled vdP systems

Here, we review a minimal nonlinear model [13] capable of describing transitions between two distinct oscillation modes in systems consisting of two symmetrically coupled self-oscillators. The motivating physical context is the Bénard–von Kármán vortex street generated in the wake of a circular cylinder.

Although the van der Pol equation is the paradigmatic model for self-oscillations *via* a Poincaré–Andronov bifurcation, it cannot capture the mirror symmetry intrinsic to vortex shedding, nor distinguish between the two physically different shedding modes: a symmetric double-row and an antisymmetric alternating pattern.

To remedy this, it is proposed [13] that vortex shedding results from the interaction of two identical nonlinear oscillators, one on each side of the cylinder, whose coupling breaks or preserves symmetry depending on physical parameters. The objective is to characterize how coupling affects the emergence and transition of oscillatory modes, and to identify the mechanisms by which symmetric and antisymmetric behaviors arise.

3.1 Linear symmetric coupling of two oscillators

Let (x_1, x_2) denote the oscillation amplitudes of the two symmetric vortex emitters. The system satisfies permutation symmetry $(x_1, x_2) \mapsto (x_2, x_1)$. Two fundamental oscillation modes are possible:

- *Symmetric mode* Θ_1: $x_1(t) = x_2(t)$ (in-phase),
- *Antisymmetric mode* Θ_2: $x_1(t) = x_2(t + T/2)$ (out-of-phase).

A simple linear model with coupling β,

$$\ddot{x}_1 + x_1 + \beta x_2 = 0, \qquad \ddot{x}_2 + x_2 + \beta x_1 = 0, \tag{6}$$

admits these two normal modes with frequencies $1 \pm \beta$. However, being Hamiltonian, it cannot generate limit cycles or stable self-sustained oscillations.

3.2 One-parameter coupled van der pol model

To introduce nonlinear self-excitation, it is proposed the system

$$\ddot{x}_1 - \varepsilon\left(1 - \left(x_1^2 + x_2^2\right)\right)\dot{x}_1 + x_1 + \beta x_2 = 0, \tag{7}$$

$$\ddot{x}_2 - \varepsilon\left(1 - \left(x_1^2 + x_2^2\right)\right)\dot{x}_2 + x_2 + \beta x_1 = 0, \tag{8}$$

which generalizes the van der Pol nonlinearity to two symmetrically coupled oscillators. For $\varepsilon > 0$ the origin becomes unstable and a stable limit cycle appears, analogous to the onset of periodic vortex shedding for Reynolds numbers $Re > Re_c$. When $\beta = 0$, the system has full $O(2)$ symmetry and supports a degenerate continuum of oscillatory states parametrized by an angle. When $\beta \neq 0$, only two periodic solutions remain:

$$\Theta_1 : x_1 = x_2, \qquad \Theta_2 : x_1 = -x_2, \tag{9}$$

with Θ_1 stable for $\beta > 0$ and Θ_2 stable for $\beta < 0$. Thus, the sign of β selects the oscillation mode.

3.3 Two-parameter coupled van der pol model

At $\beta = 0$ the symmetry is too high and the system becomes degenerate, producing an infinite family of periodic solutions. To obtain robust transitions, a second coupling parameter γ is introduced in the nonlinear dissipation:

$$\ddot{x}_1 - \varepsilon\left(1 - x_1^2 - (1 + \gamma)x_2^2\right)\dot{x}_1 + x_1 + \beta x_2 = 0. \tag{10}$$

$$\ddot{x}_2 - \varepsilon\left(1 - x_2^2 - (1+\gamma)x_1^2\right)\dot{x}_2 + x_2 + \beta x_1 = 0. \tag{11}$$

This breaks phase symmetry while preserving permutation symmetry. The enlarged (β, γ) parameter space allows non-degenerate transitions and introduces two *mixed modes*,

$$\Theta_{12}, \qquad \Theta_{21}, \tag{12}$$

intermediate between the symmetric and antisymmetric states.

3.4 Two transition scenarios

Depending on the sign of γ, two qualitatively different transition mechanisms appear.

3.4.1 Scenario I: $\gamma < 0$

Mixed modes Θ_{12} and Θ_{21} exist but are *unstable*. As β varies:

- $\beta < -c|\gamma|$: Θ_2 stable, Θ_1 unstable.
- $-c|\gamma| < \beta < +c|\gamma|$: *bistability* of Θ_1 and Θ_2.
- $\beta > c|\gamma|$: Θ_1 stable, Θ_2 unstable.

The transition occurs directly between the pure modes through a region of bistability, where c is a positive constant depending on ε.

3.4.2 Scenario II: $\gamma > 0$

Mixed modes are now *stable*. As β varies:

- $\beta < -c|\gamma|$: Θ_2 stable.
- $-c|\gamma| < \beta < c|\gamma|$: the system selects one of the mixed modes, while Θ_1 and Θ_2 are unstable.
- $\beta > c|\gamma|$: Θ_1 stable.

Thus, the transition $\Theta_2 \to \Theta_1$ occurs *through* mixed modes.

3.5 Analytical derivation of slow phase dynamics

To justify the numerical bifurcation structure, a slow-phase reduction is performed. Writing the solution as

$$x_1 = r_0(t+\psi)\cos\phi + \delta x_1, \qquad x_2 = r_0(t+\psi)\sin\phi + \delta x_2, \tag{13}$$

where $r_0\ (t)$ is the periodic orbit of the van der Pol equation, linearized equations for the perturbations are derived. A Fredholm solvability condition using the adjoint operator leads to an evolution equation for the phase difference (with k, l, s calculable functions of ε):

$$k(\varepsilon)\dot{\phi} = l(\varepsilon)\beta \cos(2\phi) + s(\varepsilon)\gamma \sin(4\phi), \tag{14}$$

whose equilibria reproduce exactly the two scenarios described above. This confirms that the transition mechanisms are structurally dictated by symmetry and coupling.

3.6 Physical interpretation and contemporary challenges

The model provides a qualitative description of the symmetry-breaking transitions between different vortex-shedding regimes [14, 15]. In fluid mechanics, the two oscillators may represent the two sides of the cylinder. Depending on coupling parameters related to geometric confinement or flow properties, the system may shed vortices symmetrically, antisymmetrically, or in mixed configurations.

The main insights of the chapter include:

- a minimal nonlinear model of two coupled van der Pol–type oscillators,
- two robust transition scenarios controlled by (β, γ),
- emergence of mixed modes absent in single-oscillator descriptions,
- analytical confirmation *via* slow-phase dynamics.

This framework may apply broadly to systems of coupled self-oscillators in physics, biology, and engineering, where symmetry and coupling shape the observed dynamics.

4. Four-dimensional models deriving from the vdP equation

Self-sustained oscillations arise in many physical, biological, and engineering systems and typically result from a balance between nonlinear amplification and dissipation. As it is known, the vdP system is a classical nonlinear oscillator exhibiting stable limit cycles for positive nonlinear damping. Motivated by the previous model of symmetric/antisymmetric Bénard–von Kármán vortex streets, here several four-dimensional or complex generalizations of vdP are recalled [16] and the central role of symmetry in organizing their dynamics is highlighted. A final model introduces a quadratic nonlinearity that breaks inversion symmetry and naturally generates a second harmonic, with additional regions of quasiperiodic and chaotic dynamics.

4.1 Four-dimensional extensions of the van der pol oscillator

The vdP oscillator $\ddot{x} - \epsilon(1 - x^2)\dot{x} + x = 0$ presents the symmetries: time translation, inversion $x \to -x$, and parameter inversion $(x, \epsilon) \to (-x(-t), -\epsilon)$. A unique limit cycle exists for all ϵ and is stable when $\epsilon > 0$. Symmetry and uniqueness imply the half-period relation $x(t + T/2) = -x(t)$.

4.1.1 Complex van der pol equation

Introducing $z = x_1 + ix_2$ yields the four-dimensional system

$$\ddot{z} - \varepsilon\left(1 - |z|^2\right)\dot{z} + z = 0, \tag{15}$$

with $U(1)$ phase symmetry $z \to e^{i\psi}z$, conjugation symmetry, and parameter inversion. For $\epsilon > 0$, solutions converge to oscillations of the form $z(t) = e^{i\phi}r(t)$, representing radial oscillations along a fixed direction in the (x_1, x_2) plane. The trivial solutions $z = e^{\pm it}$ are unstable.

4.1.2 Phase-symmetry-breaking extension

To model vortex shedding mode selection, a term $i\beta\overline{z}$ is added:

$$\ddot{z} - \varepsilon\left(1 - |z|^2\right)\dot{z} + z + i\beta\overline{z} = 0. \tag{16}$$

Rotational symmetry reduces to reflection symmetry, and two oscillatory modes appear:

- Θ_1: $x_1 = x_2$ (symmetric), stable for $\beta > 0$;
- Θ_2: $x_1 = -x_2$ (antisymmetric), stable for $\beta < 0$.

These correspond to symmetric and antisymmetric vortex streets.

4.1.3 Generalized model with conjugation-symmetry breaking

Perturbing nonlinear dissipation yields

$$\ddot{x}_1 - \varepsilon(1 - x_1^2 - (1-\gamma)x_2^2)\dot{x}_1 + x_1 + \beta x_2 = 0, \tag{17}$$

$$\ddot{x}_2 - \varepsilon(1 - x_2^2 - (1-\gamma)x_1^2)\dot{x}_2 + x_2 + \beta x_1 = 0. \tag{18}$$

Remaining symmetries include time translation, diagonal reflection, and β-sign inversion. Two nonlinear transition scenarios between Θ_1 and Θ_2 emerge:

- *Scenario I*: mixed modes are unstable;
- *Scenario II*: mixed modes are stable.

These two scenarios have been discussed in the previous section.

4.2 Model without inversion symmetry: Second-harmonic generation

Introducing a quadratic term that breaks inversion symmetry:

$$\ddot{z} - \varepsilon\left(1 - |z|^2\right)\dot{z} + z + \beta z^2 = 0. \tag{19}$$

In real variables:

$$\ddot{x}_1 - \varepsilon(1 - r^2)\dot{x}_1 + x_1 + \beta(x_1^2 - x_2^2) = 0, \tag{20}$$

$$\ddot{x}_2 - \varepsilon(1 - r^2)\dot{x}_2 + x_2 + 2\beta x_1 x_2) = 0. \tag{21}$$

Symmetries include time translation, conjugation, and parameter inversion $(z, \beta) \to (-z, -\beta)$, but *not* inversion symmetry.

4.2.1 Dynamical regions

Two parameter regions, separated by curves $\beta_1\ (\epsilon)$ and $\beta_2\ (\epsilon)$, are identified:

Region I (Second-Harmonic Generation). A unique stable limit cycle exists. Oscillations satisfy a 2:1 frequency ratio:

$$T_1 = \frac{T}{2}, \qquad T_2 = T.$$

The (x_1, x_2) projection forms an even-lobed, figure-eight pattern. This is the dynamical analogue of optical second-harmonic generation, where breaking inversion symmetry produces frequency doubling.

Region II (Quasiperiodicity and Chaos). For $\beta_1 < \beta < \beta_2$, the system exhibits:

- quasiperiodic motion on tori,
- periodic windows preserving the 2:1 frequency ratio,
- chaotic attractors with exponential sensitivity to initial conditions.

4.3 Symmetry consequences and contemporary challenges

Symmetry considerations provide a coherent framework for understanding the dynamics of all vdP-based four-dimensional models discussed [16, 17]. Phase-symmetry breaking selects between symmetric and antisymmetric oscillation modes relevant to vortex shedding. Additional symmetry-breaking terms modify the transition structure between these modes. A final model demonstrates that breaking inversion symmetry introduces natural second-harmonic generation and a narrow region of quasiperiodic and chaotic motion. These results show how simple symmetry-guided modifications of vdP-like oscillators can reproduce a wide variety of nonlinear phenomena, from optical frequency doubling to vortex dynamics [18–20].

5. Final conclusions

In this chapter, the Homotopy Analysis Method has been shown to be globally valid, and to provide high-accuracy analytical approximations for nonlinear oscillators of Liénard type. For the van der Pol system, it produces closed-form amplitude formulas with errors below 0.1% and accurate reconstructions of the limit-cycle shape. For general Liénard equations, HAM yields amplitude, frequency, and waveform expansions that agree closely with numerical results.

These advances demonstrate HAM's strength as a modern tool for nonlinear systems analysis and highlight promising research directions in mathematical dynamics and applied physics.

We have also reviewed the importance of the van der Pol oscillator as a foundational model for self-sustained oscillations across physics, engineering, and biology. Beyond the classical second-order formulation, numerous extended models have been built upon it, such as those coupled vdP systems here recalled and useful to have a first approach for the understanding of the shedding of vortices behind an obstacle. Also, other models such as the forced vdP oscillator, the van der Pol–Duffing oscillator, and networks of coupled vdP oscillators are important to explain different aspects of population dynamics and neural rhythms. In all these derived models, symmetry considerations play a crucial role in their qualitative behavior. Thus, coupled vdP networks often exhibit permutation symmetries corresponding to oscillator interchangeability, leading to synchronized, anti-phase, or clustered oscillatory states depending on the coupling topology. These symmetries govern bifurcation structure, synchronization patterns, and transitions to more complex dynamics. The study of the different dynamical regimes of this type of systems continues to be promising and leads us to contemporary challenges to be assumed and explored in the world of nonlinear systems.

Author details

Ricardo López-Ruiz
Department of Computer Science, Faculty of Science, University of Zaragoza, Zaragoza, Spain

*Address all correspondence to: rilopez@unizar.es

References

[1] Liao S. Beyond Perturbation: Introduction to the Homotopy Analysis Method. 1st ed. New York: Chapman and Hall/CRC; 2003. DOI: 10.1201/9780203491164

[2] Giacomini H, Neukirch S. Number of limit cycles of the Liénard equation. Physical Review E. 1997;**56**:3809-3813. DOI: 10.1103/PhysRevE.56.3809

[3] Llibre J, Zhang X. Limit cycles of the classical Liénard differential systems: A survey on the Lins Neto, de Melo and Pugh's conjecture. Expositiones Mathematicae. 2017;**35**:286-299. DOI: 10.1016/j.exmath.2016.12.001

[4] Abbasbandy S, Tan Y, Liao SJ. Newton-Homotopy analysis method for nonlinear equations. Applied Mathematics and Computation. 2007; **188**:1794-1800. DOI: 10.1016/j.amc.2006.11.136

[5] Abbasbandy S, López JL, López-Ruiz R. The homotopy analysis method and the Liénard equation. International Journal of Computer Mathematics. 2011; **88**:121-134. DOI: 10.1080/00207160903248659

[6] López JL, Abbasbandy S, López-Ruiz R. Formulas for the amplitude of the van der Pol limit cycle through the Homotopy analysis method. Scholarly Research Exchange. 2009;**854060**:7. DOI: 10.3814/2009/854060

[7] Cartwright ML. Van der Pol's equation for relaxation oscillation. In: Lefschetz S, editor. Contributions to the Theory of Non-linear Oscillations II. Vol. 29. Princeton Univ Press, Princeton, New Jersey: Ann of Math Studies; 1952. pp. 3-18

[8] López JL, López-Ruiz R. Approximating the amplitude and form of limit cycles in the weakly nonlinear regime of Liénard systems. Chaos, Solitons and Fractals. 2007;**34**: 1307-1317. DOI: 10.1016/j.chaos.2006.04.031

[9] López JL, López-Ruiz R. The limit cycles of Liénard equations in the strongly nonlinear regime. Chaos, Solitons and Fractals. 2000;**11**:747-756. DOI: 10.1016/S0960-0779(98)00189-1

[10] López-Ruiz R, López JL. Bifurcation curves of limit cycles in some Liénard systems. Int J Bifurcat Chaos. 2000;**10**: 971-980. DOI: 10.1142/S0218127400000694

[11] Giné J. Liénard equation and its generalizations. International Journal of Bifurcation and Chaos. 2017;**27**:1750081. DOI: 10.1142/S021812741750081X

[12] Qian X, Wang S, Yang J. Characterization of the generalized Liénard polynomial systems having special invariant algebraic curves. Qualitative Theory of Dynamical Systems. 2025;**24**:244. DOI: 10.1007/s12346-025-01404-w

[13] López-Ruiz R, Pomeau Y. Transition between two oscillation modes. Physical Review E. 1997;**55**:R3820(R). DOI: 10.1103/PhysRevE.55.R3820

[14] Zhang J, Childress S, Libchaber A, et al. Flexible filaments in a flowing soap film as a model for one-dimensional flags in a two-dimensional wind. Nature. 2000; **408**:835-839. DOI: 10.1038/35048530

[15] Eshraghi J, Rajendran LK, Yang W, et al. On flowing soap films as experimental models of 2D Navier–Stokes flows. Experiments in Fluids. 2021;**62**:162. DOI: 10.1007/s00348-021-03238-z

[16] López-Ruiz R. Symmetry induced oscillations in four-dimensional models deriving from the van der Pol equation. Chaos, Solitons and Fractals. 2004;**21**: 55-61. DOI: 10.1016/j.chaos.2003.09.033

[17] Yeşiltaş Ö. Factorization and Lie point symmetries of general Lienard-type equation in the complex plane. Physica Scripta. 2009;**80**:055003. DOI: 10.1088/0031-8949/80/05/055003

[18] Tang J, Chen Z. Amplitude control of a limit cycle in a coupled van der pol system. International Journal of Bifurcation and Chaos. 2006;**16**:487-495. DOI: 10.1142/S0218127406014952

[19] Wang Y, Li X, Shen Y. Study on mechanical vibration control of limit cycle oscillations in the van der Pol oscillator by means of nonlinear energy sink. Journal of Vibration Engineering & Technologies. 2024;**12**:811-819. DOI: 10.1007/s42417-023-00877-w

[20] Cui W, Zhao L, Ge Y, et al. A generalized van der Pol nonlinear model of vortex-induced vibrations of bridge decks with multistability. Nonlinear Dynamics. 2024;**112**:259-272. DOI: 10.1007/s11071-023-09047-9

Section 2

Applications in Biological and Physical Systems

Chapter 2

Analysis of the Behavior of Permutation Entropy and the Embedding Dimension of Takens in the Cardiac Bioelectric Signal in Healthy Subjects and Critically Ill Adult Patients

Pedro Eduardo Alvarado Rubio and
Ricardo Lino Mansilla Corona

Abstract

Biological systems contain a complex structure, varying in size and scale over time, self-organizing and interacting with their environment. How they optimize robustness, resilience and fragility, functioning in an interconnected sequence of processes-products-outcomes, interacting with the hierarchically organized constituent subsystems, is unknown. We use entropy and Embedding Dimension (ED) as tools to analyze the cardiac bioelectrical signal (CBES) in critically ill patients, contrasting them with those of healthy individuals. Based on thermodynamics, information theory and nonlinear dynamics, emphasizing entropy in particular, permutation entropy (PE) is used to quantify the loss of information or complexity of physiological systems. The representation of these magnitudes is essential to adequately characterize the process under study and relevant components. By reconstructing the phase space of cardiac bioelectrical dynamics using Takens' theorem and applying techniques such as mutual information and correlation integral, the study identifies decreases in ED and increases in PE often coinciding with patient deterioration. Conversely, increases in ED and reductions in PE suggest improvement or stability. We used a Holter monitor to obtain time series from critically ill patients and healthy subjects to characterize the dynamics of their cardiac signals. The analysis revealed that critically ill patients frequently presented ED values dropping to 4, often before death; healthy patients typically maintained ED values between 5.8 and 6.7, with PE values in narrower ranges.

Keywords: critically ill patients, entropy, permutation entropy (PE), embedding dimension (ED), cardiac bioelectrical signal (CBES), time series and Takens' theorem

IntechOpen

1. Introduction

Information and Cybernetics Theory [1–4] and complex systems have allowed us to observe and study other aspects of bioelectric signals beyond the perception of our senses and to better understand their dynamic behavior. Bioelectric signals such as electroencephalograms and electrocardiograms show that bioelectric activity has the characteristic of being unique to everyone [5].

For the functioning of organs, which are formed by cell conglomerates interacting at various levels with other bidirectional and/or multidirectional communication, in biological organisms such as the human being, the signals collected by analog clinic equipment have been very useful in the research to measure biological rhythms that characterize the performance of our organs, for example: electrocardiography. The processing of these signals is one of the scientific and technical areas of greater development in biomedical engineering with a very important impact on current clinical practice [6].

Rudolf Julius Emmanuel Clausius introduced the concept of entropy in 1865 to express the degradation of energy in industrial thermodynamics. The word "entropy" derives from the Greek "in" (inside) and "trope" (change or transformation). It was originally conceived to describe the dispersion of energy within a thermodynamic system. Clausius used this concept to explain the irreversibility of natural processes and the eventual tendency toward thermal equilibrium [7].

Subsequently, Ludwig Boltzmann, James Clerk Maxwell and Josiah Willard Gibbs developed the statistical mechanical theory of thermodynamics, providing the possibility of measuring the uncertainty of a physical system, that is, the unpredictability of a microscopic system with unknown degrees of freedom, to an observer of a macroscopic system [8]. This concept of entropy has played a decisive role in various branches of knowledge, especially in physics and information theory, significantly evolving for almost 160 years, until today.

From the thermodynamic point of view, living biological organisms have a functional order very different from that of a machine that by means of heat produces work as a result; in this process, the degradation of energy that is not used to generate work is called entropy [7]. Ilya Prigogine considered that systems far from equilibrium could be a source of order and approach to irreversible processes of a new type of dynamic states of matter; based on the thermodynamic theory of structures, he called them "dissipative structures," which generate coherent systems of oscillatory enzymes as a product of reactive collisions [9]. For example, Krebs cycle [10].

In the 1940s, Claude Elwood Shannon established the foundations of a new scientific branch based on that magnitude called Entropy; the "Information Theory," which argues that, to any signal (measurement or random variable) an entropy value can be assigned, that is to say, measure its unpredictability and, consequently, the information content of that physical system, also being useful for measuring the variability of biological systems. Consequently, it also measures the amount of information and the freedom of the system [11]. This point of view provides a coherent interpretation of the various entropy formulas and many experimental facts, such as the self-organized structures of closed or open systems (but not of isolated systems). In an open system, there is a bidirectional flow with mass, energy, entropy, information, and meaning exchanged with its environment, allowing the system to remain far from the thermodynamic equilibrium [12].

A characteristic of entropy is characterized by is its extensive and not preserved character, linked to the probabilistic characteristics of the accessible microstates of a system, in the corresponding phase space. Other rates of entropy variation over time are possible when phase space geometry is fractal [13], such as the temporal evolution of the cardiac bioelectric signal (SBEC), having fractal behavior, as a non-harmonic oscillator [14].

The number of essential variables to fully characterize the behavior of a dynamic system is fundamental, since it indicates the minimum number of dimensions that must be considered for a complete understanding of the phenomenon. The results obtained with analog clinical devices are responsible for the measurements; These are observable macroscopic systems under study that do not reveal explicit information about the number of magnitudes necessary for the complete description of the underlying dynamic system. For this reason, it is necessary to use immersion techniques [15], to rebuild the physical phase space of the system from the observables from the medical equipment.

The variable of interest in this study is the so-called PE [16], which provides an appropriate characterization of the complexity of time series through ordinal patterns defined by the relations between numerical values, allowing us to understand complex and chaotic systems. PE has the enormous advantage of having notable applications to biomedical problems [17–21].

This study demonstrates that the entropy values, as well as the embedding dimension, significantly differ between healthy individuals and critically ill adult patients.

2. Mathematical structure

Consider a dynamic system given by diffeomorphism $\varphi{:}M \rightarrow M$ on a compact variety M that we will suppose soft, $\varphi \epsilon\ Diff^{d}\ (M)$. Dynamic system trajectories are built through $x_t = \varphi^t\ (x_0)$. It is pertinent to remember that $\{\varphi^t\}_{t\in\mathbb{R}}$ is an uniparametric group of diffeomorphisms. It is also assumed that there is an observable $y_t = g(x_t)$, where the function $g{:}M \rightarrow \mathbb{R}, g \in C^l(M, \mathbb{R})$. Then, the following result is fulfilled:

Theorem (Takens) [15]: For a compact variety M of dimension $m, l \geq 1$ and $n > 2\,m$ there is an open and dense subset $U \in Diff^{\,l}(M) \times C^l(M)$ such that for any $(\varphi,g) \in U$, the following application is an immersion of M en $\mathbb{R}^n$:

$$M \ni x \mapsto \left(g(x), g(\varphi(x)), \dots, g\left(\varphi^{n-1}(x)\right)\right) \in \mathbb{R}^n$$

Note that the conclusion of the theorem is generic in (φ, g).

This means that using the observable macroscopic that provides the time series $\{x_t\}$, the dynamics of the cardiac signal can be fully recovered from building vectors of the type $\left(x_t, x_{t+\tau}, \dots, x_{t+(n-1)\tau}\right)$ which generate the attractor of the dynamic system associated with the cardiac signal [22, 23]. The parameters τ and n must be properly chosen. The first is known as delay and the second as an embedding dimension (ED). **Figure 1** describes the embedding process. The phenomenon under study occurs in a differential variety (in our case the surface of the heart muscle) of which we only have an observable macroscopic (which is the EKG) from which we generate an object O, constituted by the vectors $\left(x_t, x_{t+\tau}, \dots, x_{t+(n-1)\tau}\right)$ in some space of the type $\mathbb{R}^n$. This object O retains all the topological properties of the phenomenon under study, particularly the topological, fractal dimensions and certain complexity measures.

The previously described method guarantees the existence of parameters τ and n, however, does not provide a procedure for obtaining. Next, we will describe what are

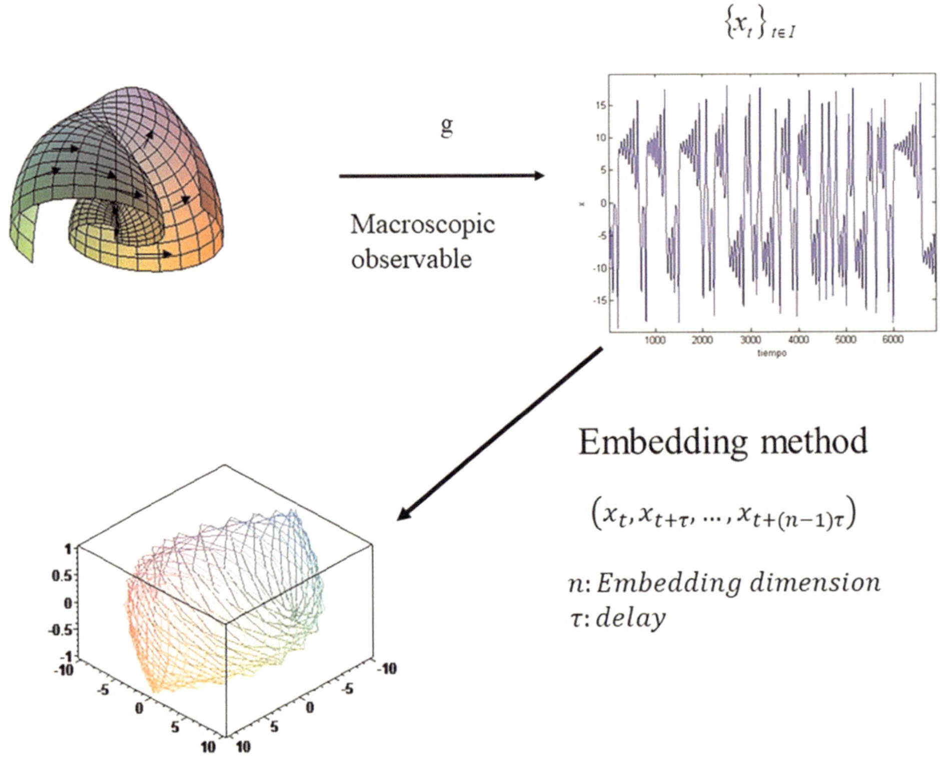

Figure 1.
Embedding process in time series [24, 25].

the most common procedures to estimate these magnitudes. The basic theoretical concepts in this estimate task are the function of mutual information and the correlation integral.

Let $\{x_t\}$, a series of time, we call mutual information to the following function:

$$M(\tau) = \sum_{t=1}^{N-\tau} P(x_t, x_{t+\tau}) \, ln \left[\frac{P(x_t, x_{t+\tau})}{P(x_t)P(x_{t+\tau})} \right]$$

Where:

$P(x_t, x_{t+\tau})$: is the function of joint probability density between x_t y $x_{t+\tau}$.

$P(x_t)$: It is the probability density function of the signal x_t.

Is known that the correct value of τ is the first local minimum of the mutual information function [26, 27].

The criterion to find the correct ED depends on the construction called Correlation Integral:

Let $\{X_i\}$ the collection of vectors is built as follows: $X_i = \left(x_{t_i}, x_{t_i+\tau}, \dots, x_{t_i+(n-1)\tau}\right)$, $t_i \leq N$. We call comprehensive correlation of dimension n a:

$$C_n(\varepsilon) = \binom{n}{2}^{-1} \sum_{i<j} \mathbb{H}(\|X_i - H_j\| < \varepsilon)$$

where $\mathbb{H}$ is the function of Heaviside and $\|.\|$ is the Euclidean norm of $\mathbb{R}^n$. With this, the correlation dimension is defined:

$$d_c = \lim_{\varepsilon \to 0} \lim_{n \to +\infty} \frac{\log C_n(\varepsilon)}{\log \varepsilon}$$

Figure 2 shows the graphics of the correlation integrals for different values of the ED. Note that there is a perceptible separation between the graphics of the correlation integrals for dimensions 6 and 7. From the integral correlation graph for dimension 7, the remaining graphs (dimensions 8 to 15) appear closer than the rest. ED is obtained, following the criteria mentioned above, so the appropriate ED is 7. It is pertinent to add that the EKG, with which the correlation integrals shown in **Figure 2**, belonged to a 24-year-old healthy man; however, it is important to mention that later, also in healthy subjects some show an ED of up to 8.

Basically, the correlation dimension is the slope of the correlation integral at the origin. The criterion for selecting an ED is as follows [22, 25, 28].

"Choose increasing ED and in each case, calculate the correlation integral. When no changes in the behavior of the correlation integral with respect to the increase in the ED are observed, then an appropriate ED will have been found."

Its definition is as follows: Let $\{x_1, \dots, x_N\}$ a series of time and be m < N. Let S_m be the order permutation group m and for all $0 < t \le N - m + 1$ the mobile window $x_m(t) = (x_t, x_{t+1}, \dots, x_{t+m-1})$.. We say that this window is of type $\pi_i \in S_m$ si $\pi_i = (i_1, \dots, i_m)$ is the only permutation that satisfies the following conditions:

a. $x_{t+i_s} \le x_{t+i_{s+1}}$ para todo $s = 1, \dots, m-1$

b. $i_s < i_{s+1}$ si $x_{t+i_s} = x_{t+i_{s+1}}$

Note that the application $\Phi : W^x_{t,m} S_m$ defined by $\Phi(x_m(t)) = \pi_t$ is neither injective nor surjective.

by $\pi \in S_m$ φ (x_m (t)) = π_t is neither injective nor objective.

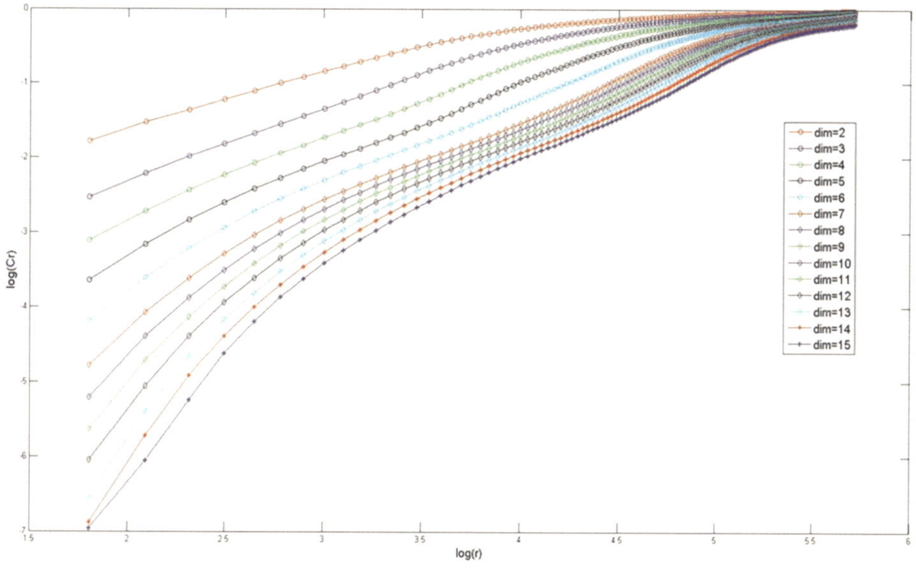

Figure 2.
Correlation integrals for values of the embedding dimension between 2 and 15 [24].

For each $\pi \in S_m$ it can be defined:

$$p_m(\pi) = \frac{\#\{t | t \leq N - m + 1, \Phi(x_m(t)) = \pi\}}{N - m + 1}$$

In this way, the permutation entropy for the time series $\{x_1, \dots, x_N\}$ with window size m is defined as:

$$PE(m) = -\sum_{\pi \in S_m} p_m(\pi) \log p_m(\pi)$$

3. Method

We studied the cardiac bioelectrical signal (CBES), assuming that its evolution is governed by a dynamic system. A three-channel Holter monitor (Scott Care Corporation. Chroma: model RZ153C) was used to monitor 15 patients, obtaining numerical data files of between 800,000 and 1,200,000 rows (depending on the subject's heart rate under study) and three columns corresponding to each of the observables. Of these, only one was used for the study by obtaining the comma-separated values (CSV) files, which makes up the time series analyzed in two groups: one of 15 critically ill patients. Of the 15 critically ill patients, only 14 are shown in the description table (**Table 1**); patient 15 is analyzed separately due to the singular behavior of its evolution.

The diagnosis of the patients was performed from the clinical point of view and was confirmed through image studies such as the computed axial tomography of the topologically involved site, skull, thorax or abdomen. Some of the patients underwent surgically intervened on one or more occasions. Chronic-degenerative associated diseases such as systemic blood pressure (SBP) were common. Holter monitoring was performed in seriously ill patients at position supine for a period of 1 h every 24 hrs from admission to his death. All Holter records in severe patients were performed with sedated patients, with amines, and ventilatory mechanical support. All patients were evaluated and scored using **APACHE II** (**Table 2**) [32].

No	Age	Genus	Suffering	Surgery	APACHE	Outcome	Images	Holters
1	38	M	FLS by Tumor	Yes	29/60%	Died	CCTS	1
2	65	F	SAH SAH + P	Yes	27/70%	Died	CCTS	4
3	60	M	SAP	Yes	32/75%	Died	ACTSIVC	1
4	64	M	PCH	Yes	27/55%	Died	CCTS	6
5	69	F	Fisher III SAB + PE + CATC	Yes	26/60%	Died	CCTS +TCT	3
6	55	F	PS CAVM	Yes	30/70%	Died	CCTS +TCT	3
7	47	F	Fisher IV BBA and SAH	Yes	32/75%	Died	CCTS	4
8	33	F	Fisher IV SAH by SAH	Yes	29/55%	Died	CCTS	16

N o	Age	Genus	Suffering	Surgery	APACHE	Outcome	Images	Holters
9	56	M	ASH	Yes	32/70%	Died	CCTS	2
10	70	F	HCS	No	30/75%	Died	CCTS	3
11	52	M	CIRS with H	Yes	30/55%	Died	CCTS	2
12	61	F	Fisher IV SAH by SAH + DM	Yes	28/60%	Died	CCTS	2
13	78	F	Fisher IV BBA	Yes	30/70%	Died	CCTS	1
14	42	F	BD	Yes	30/70%	Died	CCTS	3
			Average age 56.4 ± 12.9	APACHE: 29.1 ± 2.5/ 63.9% ± 10.59%				

*Abbreviations for diseases and complications of critically ill patients in **Table 1**.*
1. FLS by T = Frontal Lobe Syndrome by tumor. 2. SAH SAH + P = Subarachnoid Hemorrhage secondary to Systemic Arterial.
Hypertension + Pneumonia. 3. SAP = Severe Acute Pancreatitis. 4. PCH = Parenchymal Cerebral Hemorrhage. 5. Fisher III SAH + PE = Subarachnoid hemorrhage + Pulmonary Embolism + CATC = Computed Angio tomography of the Chest. 6. PS CAVM = Post Surgical of Cerebral Arterio Venous Malformation. 7. Fisher IV BBA, SAH = Fisher IV Broken Brain Aneurysm and Subarachnoid Hemorrhage. 8. Fisher IV SAH by SAH = Subarachnoid Hemorrhage by Systemic Arterial Hypertension. 9. ASH = Acute Subdural Hematoma. 10. HCS = Hemorrhage of Cerebral Stem. 11. CIRSH = Cerebellar infarction on the right side with hemorrhagic. 12. Fisher IV SAH to SAH + DM = Fisher IV subarachnoid hemorrhage, systemic arterial hypertension + type 2 diabetes mellitus. 13. Fisher IV BBA = Fisher IV Broken Brain Aneurysm. 14. BD = Brain Death (with diagnostic protocol of Brain Death [29]).
Fisher classification. The relationship between cerebral vasospasm and subarachnoid hemorrhage [30].
All patients with craniocerebral injury were evaluated using the Glasgow Coma Scale [31].
ACTSIVC = Abdominal computed tomography scan with intravenous contrast.
CCTS = Cranial Computed Tomography Scan.

Table 1.
Brief description of the patients under study.

Physiology Variable		+4	+3	+2	+1	0	+1	+2	+3	+4
Temperature – rectal (°C)		241	39–40.9		38.5–38.9	36–38.4	34–35.9	32–33.9	30–31.9	≤ 29.9
Mean arterial pressure (mm Hg)		≥ 160	130–159	110–129		70–109		5–69		≤ 49
Heart rate		≥ 180	140–179	110–139		70–109		55–69	40–54	≤ 39
Respiratory rate (nonventilated or ventilated)		≥ 50	35–49		25–34	12–24	10–11	6–9		≤ 5
Oxygenation (mmHg) a. FiO2 > 0.5 use A-aDO2	A	≥500	350–499	200–159		<200				
a. FiO2 < 0.5 use PaO2	B					>70	61–70		55–60	<55
Arterial pH		≥ 7.7	7.6–7.69		7.5–7.59	7.33–7.49		7.25–7.32	7.15–7.24	<7.15
Serum sodium (mmol/1)		> – 180	160–179	155–159	150–154	130–149		120–129	111–119	≤ 110

Physiology Variable	+4	+3	+2	+1	0	+1	+2	+3	+4
Serum creatinine (mg/dl, double point score for acute renal Failure)	>– 3.5	2–3.4	1.5–1.9		0.6–1.4		<0.6		
Hematocritto (%)	≥ 60		50–59.9	46–49.9	30–45.9		20–29.9		<20
White blood count (in 1000 mm^2)	≥ 40		20–39.9	15–19.9	3–14.9		1–2.9		<1
Glasgow-Coma-scale (GCS)	Score = 15 minus actual OCS								
Serum HCO3 (Venous, mmol/1, use if no ABG3)	≥ 52	41–51.9		32–40.9	22–31.9		18.21.9	15–17.9	<15
A = Total Acute Physiology Score APS	Sum of the 12 individual variable points								
B = Age Points	C = Chronic Health Points								
≤ 44 Years 0 Points 45.54 Years 2 Points 55–64 Years 3 Points 65–74 Years 5 Points ≥ 75 Years 6 Points	If the patient has a history of severe organ system insufficiency or is immunocompromised, assign points as follows: a. For nonoperative or emergency postoperative patients –5 points b. For elective postoperative patients –2 points								
APACHE II SCOR = Sum of a (APS Points) + B (age points) + C (Chronic Health points).									

Table 2.
The APACHE II. Severity of Disease Classification System.

N°	1	2	3	4	5	6	7	8	9	10
Sex	F	M	F	F	F	M	M	F	F	M
Age	27	26	31	30	31	33	29	32	28	30
Average age 29.7 ± 2.21										

Table 3.
Healthy young adults.

The second analysis group is made up of healthy young adults, 6 females and 4 men. We were monitored for 1 hour of recording in the morning in supine position sleeping (**Table 3**).

4. The APACHE II: Severity of disease classification system

Below are the graphs for each critically Ill patients showing the electrocardiographic (ECG) condensation of the Holter recording (in blue), from which the time series are obtained. Below are both magnitudes (permutation entropy and embedding dimension) for each patient, with their behavioral trends indicated by red arrows.

4.1 Patient 1

38-year-old male. Occasional alcoholism. Current condition begins 45 days before with changes in the behavior detected by his wife, which increases over time, adding headache. Enter Neurosurgery with suspected left frontoparietal abscess, in the trans surgical, a tumor is observed, biopsy is taken. In the post-surgical, severe face-caudal deterioration. Decompressive craniectomy is performed. Observed pulse absence in the left cerebral hemisphere (**Figure 3** (graph and images 1 and 2)). During surgery, it had output of brain matter. The patient died the same day, 2 hours after Holter's recording. The histopathological report was grade IV glioblastoma multiforme.

The graph shows an increase in PE from 3.3 to 3.9. The initial ED was 4, with brief increases to 5 and ending at 4. This behavior is consistent with the patient's severity.

4.2 Patient 2

65-year-old female with a history of systemic arterial hypertension under treatment with Telmisartan. It is brought to adult emergencies, for loss of the state of consciousness. A cranial CT scan shows subarachnoid hemorrhage (**Figure 4** (graphs 1, 3, and 5, and images 1and 2)) Fisher IV [30]. Back to the Emergency Department presents cardiorespiratory arrest. Following admission to Critical Care, the patient died 5 days later.

Graph 1, the PE is 5.5 and ED is 4. The PE subsequently increases to 5.9 and decreases to 5.7, with the ED increasing from 4 to 6.

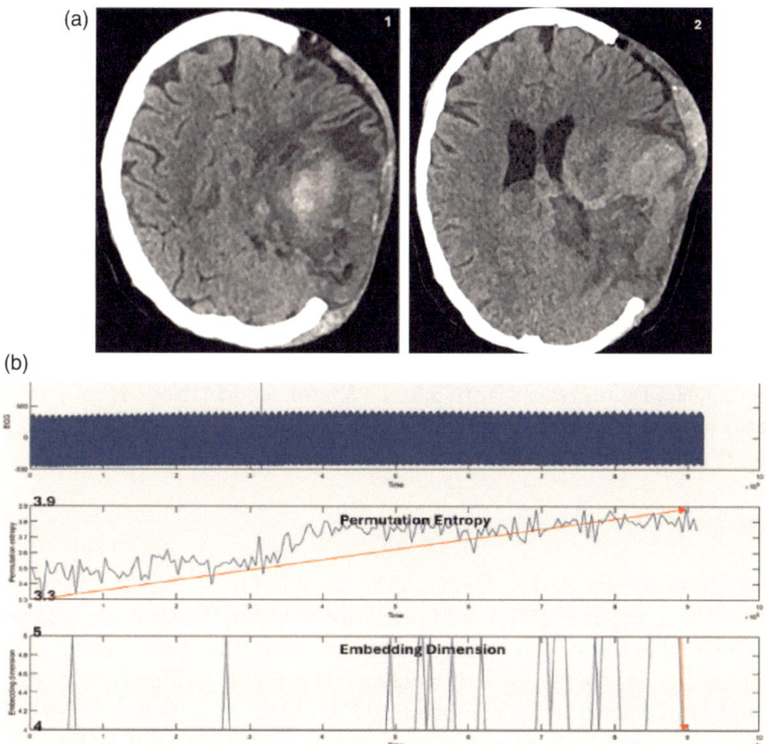

Figure 3.
(a) A cranial CT scan shows grade IV glioblastoma multiforme (images 1, 2). (b) Graph of permutation entropy and embedding dimension.

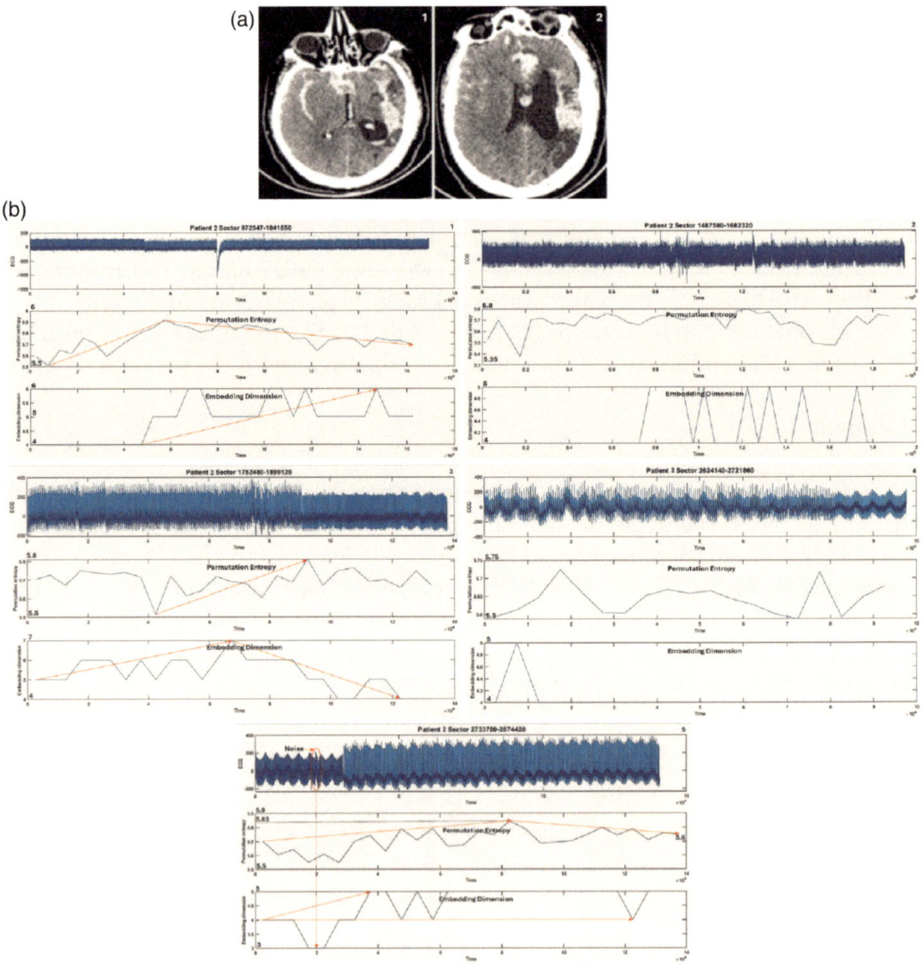

Figure 4.
(a) A cranial CT scan shows subarachnoid hemorrhage (images 1, 2). (b) Graphics of permutation entropy and embedding dimension 1-5.

In Graph 3, the PE increases from 5.5 to 5.8, and the ED drops from 7 to 4.

In Graph 5, a disturbance in the CBES condensate is observed, which may explain the drop in the ED to 3.

4.3 Patient 3

60-year-old male transferred from another hospital unit with diagnosis of biliary pancreatitis. The computed abdomen tomography was performed with intravenous contrast. The images obtained correspond to 3 coronal cuts at the height of the pancreas. Number 1 shows a normal pancreas, to compare ((**Figure 5**) graph and images 2 and 3) of the seriously compromised pancreas, these images show, increase the size of the pancreas and extensive areas of hypoperfusion of the organ, suggestive of necrosis. The patient died 3 hours after his entry to critical care unit.

It is observed that the PE has an average behavior of 5.7, and the ED is practically at 4 throughout the entire record.

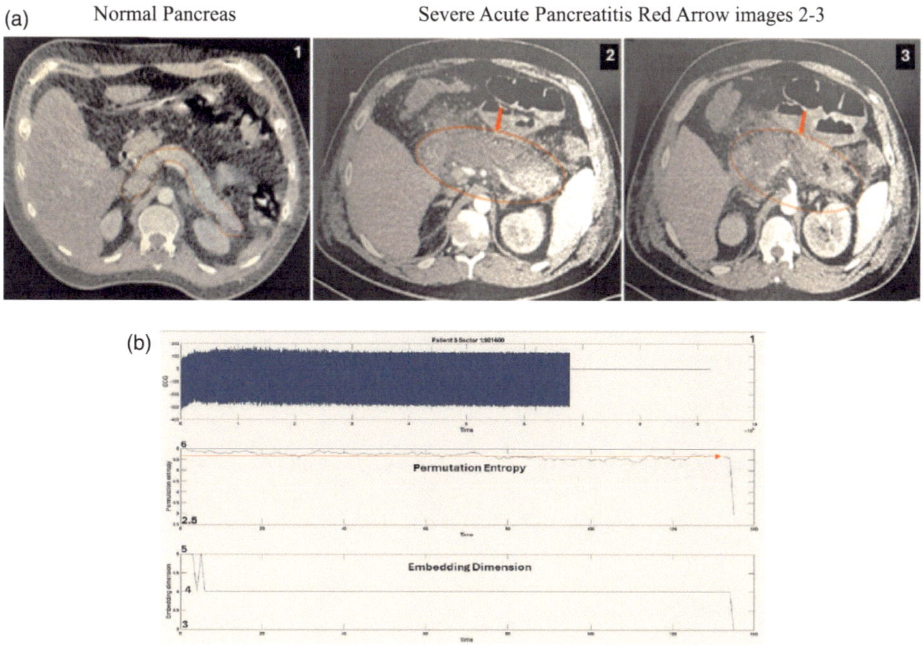

Figure 5.
(a) A cranial CT scan shows severe acute pancreatitis (images 2, 3). (b) Graph of permutation entropy and embedding dimension.

4.4 Patient 4

64-year-old man, who is found at his home with right parietal contusion in ethyl state, sleepy with nausea and vomiting, without sphincter control. Skull tomography shows multiple bruises and severe cerebral edema as shown in **Figure 6** (graphs 1–3 and images 1and 2). Craniectomy and drainage of right subdural hematoma were performed (images 3 and 4). It evolves to multiple organic failure and dies 8 days after its entry into Critical Care.

In graph 1, the PE from 5.65 drops to 5.5 and the ED increases from 4 to 7.

In graph 2, the PE increases from 5.1 to 5.6 and the ED decreases from 7 to 4.

In graph 3, the PE increases from 4.25 to 5.5 and the ED decreases from 7 to 4.

4.5 Patient 5

60-year -old female patient, who suffers from long -term systemic arterial hypertension, with a bad attachment to treatment. She is taken to the hospital, unconscious with 6 neurological evaluation points on the Glasgow scale [33]. Tomography showed subarachnoid bleeding (Fisher IV) ((**Figure 7**) (graphs 1, 3, and 4, and images 1, 2 and 3)). She assisted in ventilation in a mechanical way and subsequently entered critical care. In its evolution with sudden hemodynamic instability secondary to pulmonary embolism, dying 6 days after admission.

In graph 1, the PE ranges between 4.3 and 4.75 and the tonic ED at 4.

In graph 3, the PE of 4 increases to 5.5 and the tonic ED to 5.

In graph 4, the PE of 5.4 decreases to 4.4 and the ED of 6 decreases to 4 (**Figure 8**).

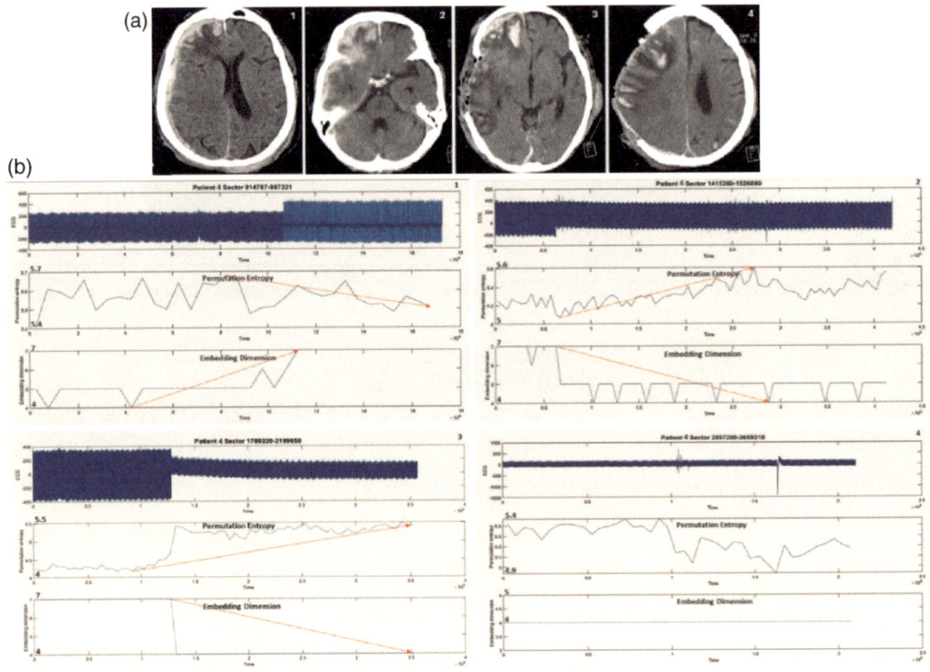

Figure 6.
(a) Severe cerebral edema (images 1, 2). Drainage of right subdural hematoma (images 3, 4). (b) Graphics of permutation entropy and embedding dimension 1-4.

4.6 Patient 6

55-year-old female with 3-year-old diabetes of evolution treated with metformin. Abdominoplasty 20 years ago, allergic to Penicillin and Sulfamethoxazole, hypertriglyceridemia of 5 years of evolution treated with Pravastatin. His current condition began 2 years ago with an increase in volume on the right side of the neck, and his behavior was cyclical and with improvement also cyclical without any particular attention. A cranial and neck CT scan showed occipital arteriovenous malformation of the right side (**Figure 9** (graphs 4 and Image 1)). During the surgical resection of malformation, it had a bleeding of 3500 cc. She was admitted to the critical care unit, to treat hypovolemia and ventilatory mechanical assistance. CT of control shows post-surgical bleeding in ventricles (Images 2, 3 and 4), draining through the Foramen of Luschka and Magendie, with partial improvement. She dies 20 days later.

In the 4 graphs, the PE only shows brief increases, with an average of 4.7, the ED is striking for its brief and frequent fluctuations, rising its ED from 4 to 5 and even in graph 3 to 7.

4.7 Patient 7

47-year-old female with type 2 diabetes for 7 years, diagnosed with subarachnoid hemorrhage. In the Emergencies, Glasgow is in 6 points, which requires advanced airway management. A cranial CT scan showed suggestive hyperdense images of subarachnoid hemorrhage with irruption to lateral ventricles and third ventricle

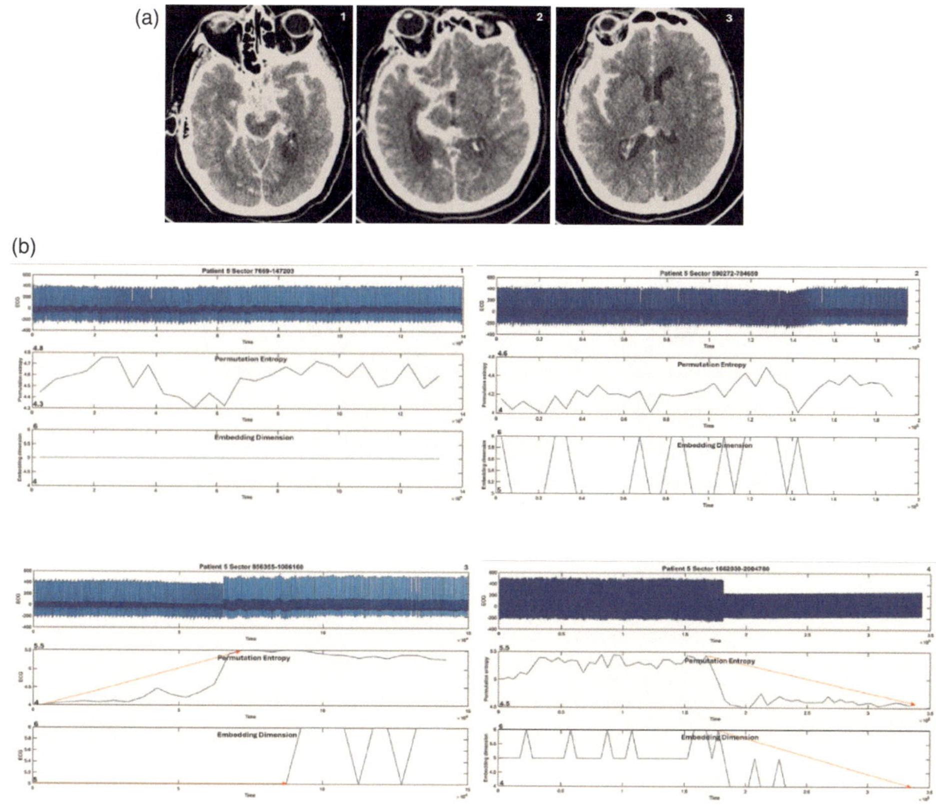

Figure 7.
(a) Subarachnoid bleeding (Fisher IV) (images 1-3). (b) Graphics of permutation entropy and embedding dimension 1-4.

(**Figure 10** (graphs 2 and Images 1, 2 and 3)). Evaluated by neurosurgery, a right-sided precoronal ventriculostomy was performed. She was admitted in critical care with a 3-point Glasgow. Its evolution with neurological deterioration in cascade, without improvement, dying 4 days after admission.

In graph 2, it is important to mention the disturbance recorded in the EKG condensate, which correlates with a false ED in 3.

4.8 Patient 8

33-year-old female. Both parents with systemic arterial hypertension, their maternal grandmother with cerebral infarction. She is carrier of systemic arterial hypertension since the age of 15 in control with 10mgrs oral via of Enalapril every 8 hrs and Metoprolol 100 mgs oral *via* every 24 hrs. Cesarean section was done 8 years ago by Preeclampsia. In consultation with a doctor, its blood pressure was of 220/140 mmHg, and left chest member paresia for which, it is referred to the Emergency Department, where it presents emesis of gastric content and increase in paresia and adding dysarthria and facial paralysis. The initial Glasgow Coma scale is of 13 points. Computed axial tomography initial of skull (**Figure 11** (graphs and Image 1)) showed hyperdense lesion in cerebral parenchyma at the level of rights basal cores and greater neurological deterioration manifesting anisocoria with right mydriasis and left miosis. The

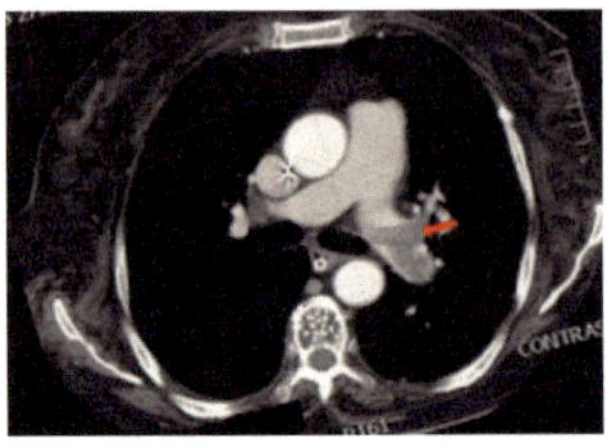

Figure 8.
Computed Angio Tomography of the chest, with pulmonary embolism (red arrow).

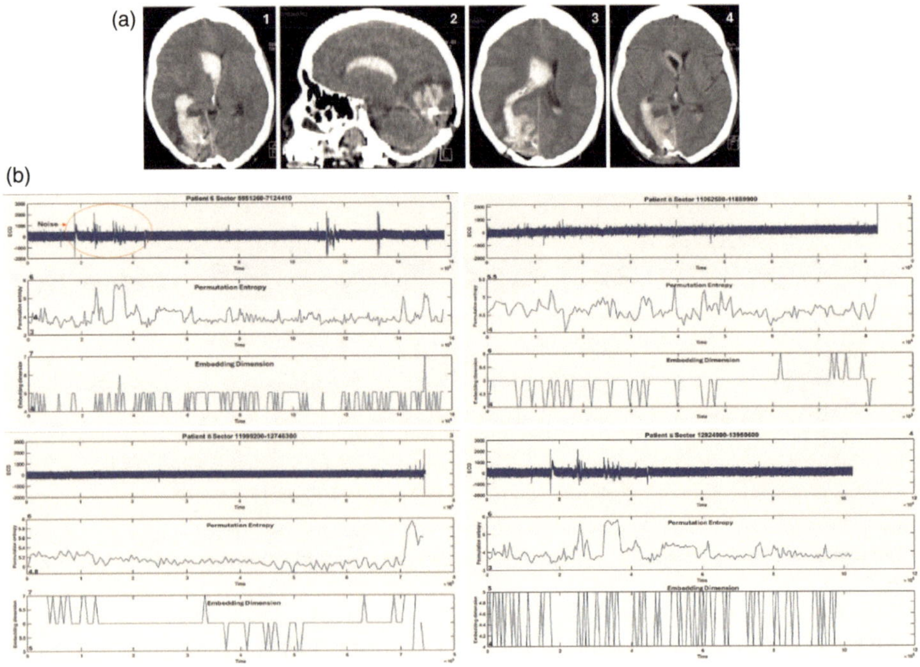

Figure 9.
(a) CT scan showed occipital arteriovenous malformation of the right side (image 1). CT scan of control shows post-surgical bleeding in ventricles (images 2-4). (b) Graphics of permutation entropy and embedding dimension 1-4.

Glasgow coma scale is of 5, and blood pressure is 240/170 mmHg. The airway is approached and sedated and assisted mechanical ventilation begins, subsequently to the operating room to perform decompressive craniectomy and hematoma drainage with bleeding (Images 2 and 3) in the surgical act of 1200 ml. The patient is admitted to the Adult Intensive Therapy Unit in the immediate post-surgical. In its evolution it is complicated with multibacterial brain abscess (Images 4 and 5), dying 31 days after admission.

In graphs 3, 5, 8, and 10, their dynamics behave by decreasing PE and increasing ED, and in graphs 6 and 9 by increasing PE and decreasing ED.

4.9 Patient 9

56-year-old male, who is found at his home with left parietal contusion in ethyl state, sleepy with nausea and vomiting, without sphincter control. A cranial CT scan

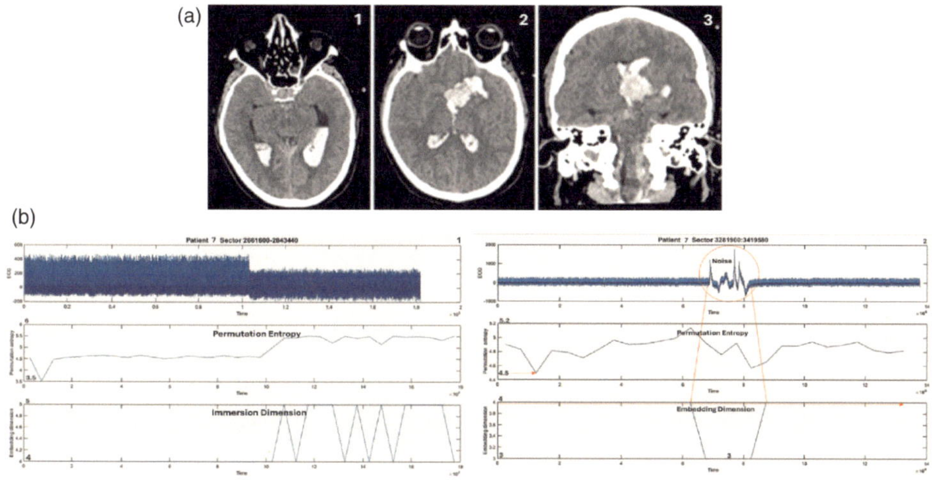

Figure 10.
(a) A cranial CT scan showed subarachnoid hemorrhage with irruption to lateral ventricles and third ventricle (images 1-3). (b) Graphics of permutation entropy and embedding dimension 1-2.

shows left hemisphere hypoperfusion and severe cerebral edema. Craniectomy and drainage of left subdural hematoma were performed (**Figure 12** (graphs and Images 1 and 2)). It evolves into multiple organic failure and dies 8 days after its entry into Critical Care.

Graph 3, its dynamics behave, decreasing the PE and the ED increases in graph 4, increasing the PE and decreasing the ED.

4.10 Patient 10

70-year-old female, type 2 diabetic and hyper systemic blood pressure, both long evolution (20 years). From its entrance to the hospital, unconscious and with ventilatory mechanical assistance, a cranial CT scan shows generalized severe cerebral edema and Fisher IV subarachnoid hemorrhage (**Figure 13** (graphs and images 1, 2 and 3)). The patient dies 6 days after admission to the critical care unit.

In graph 2, its dynamics can be observed, increasing the PE and decreasing the ED.

4.11 Patient 11

52-year-old male begins with vertigo and difficulty articulating words. A cranial CT scan shows cerebellar infarction (**Figure 14** (graphs and image 1)) that evolves too hemorrhagic. Neurosurgery places a ventriculostomy catheter by dilation of the ventricular system and subsequently performing suboccipital craniectomy (image 2). Enter the Critical Care Unit with sedation and ventilatory mechanical support. During its evolution, it was necessary to replace the ventriculostomy catheter. During this surgical act, it presents ventricular fibrillation, without response to the usual resuscitation maneuvers.

In graph 2, a tonic behavior can be observed in its dynamics in 5 of the ED.

(a)

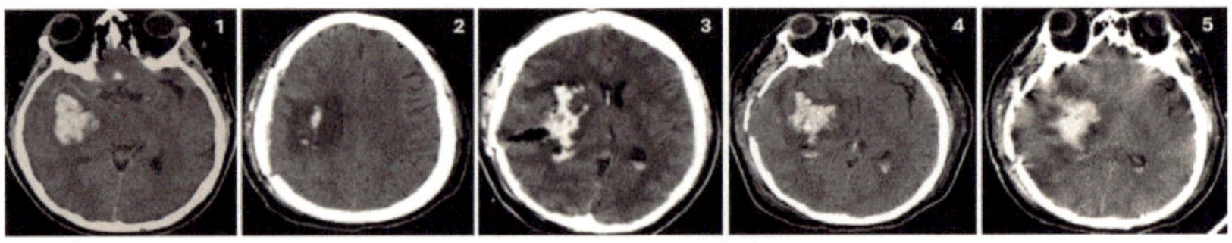

(b)

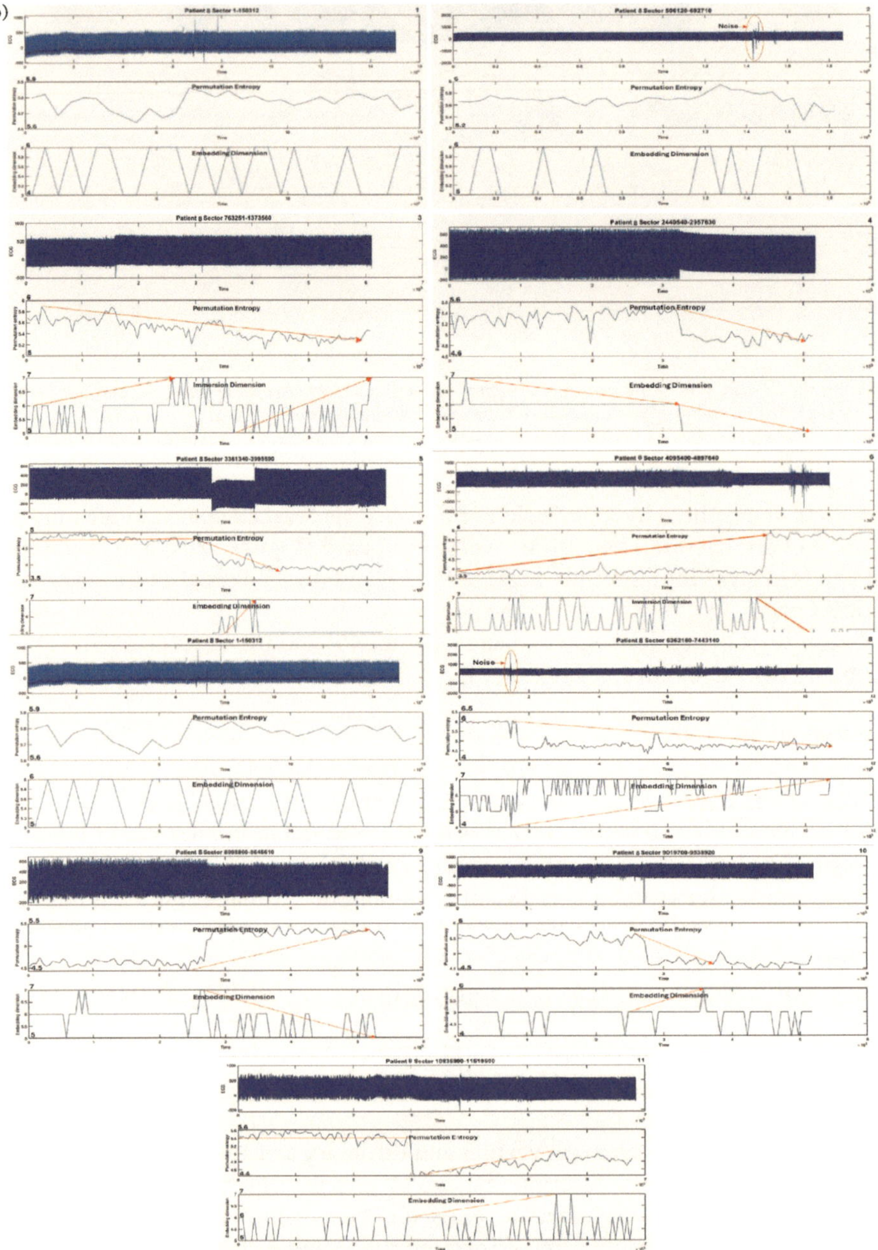

Figure 11.
(a) A cranial CT scan showed hyperdense lesion in cerebral parenchyma (image 1), subsequently decompressive craniectomy and hematoma drainage (images 2 and 3). Finally, complicated with multibacterial brain abscess (images 4 and 5). (b) Graphics of permutation entropy and embedding dimension 1-11.

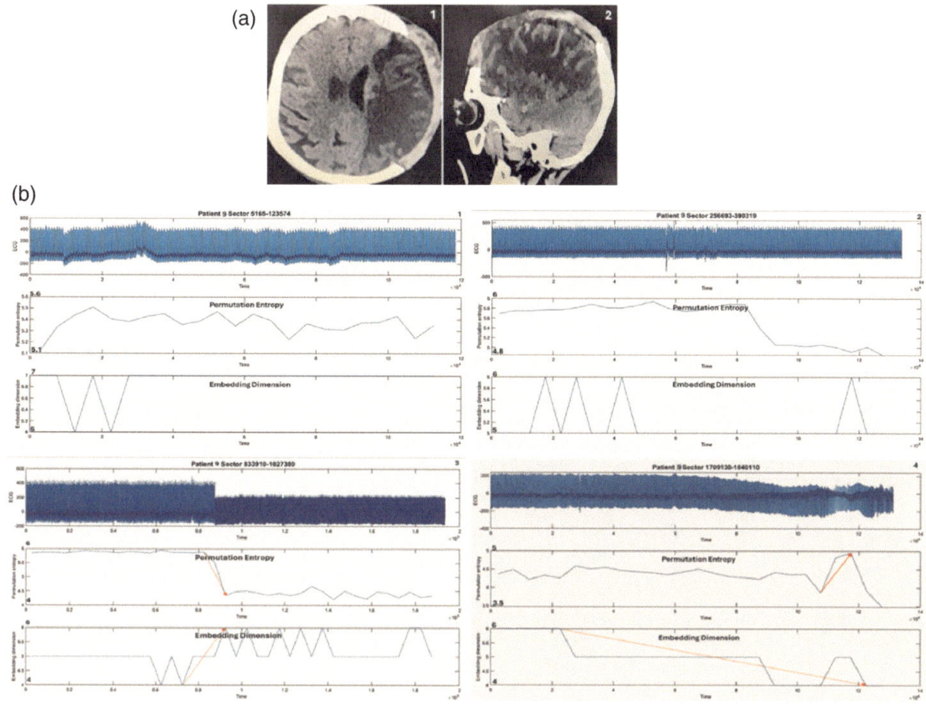

Figure 12.
(a) A cranial CT scan shows left hemisphere hypoperfusion and severe cerebral edema (image 1). Later craniectomy and drainage of left subdural hematoma (image 2). (b) Graphics of permutation entropy and embedding dimension 1-4.

4.12 Patient 12

61-year-old female, with type II diabetes with systemic 11-year systemic arterial hypertension. It was found at home lying on the floor, with cranial contusion on the right side. Enter after being treated in another hospital unit. In the Emergency Department, the CT scan showed the cerebral contusion of the right side with suggestive areas of bleeding and severe cerebral edema (**Figure 15** (graphs and images 1 and 2)). Entrance to neurosurgery for decompressive craniectomy. The patient was admitted to Critical Care for their attention, dying 7 days after its admission.

In graph 2, the paradoxical behavior of PE and ED can be observed in its dynamics, since both magnitudes decrease.

4.13 Patient 13

78-year-old female, with systemic arterial hypertension of long evolution, is found at home in an armchair without response to stimuli, so she is brought to the emergency service with 5-point Glasgow. The CT scan shows subarachnoid hemorrhage, suggestive aneurysm rupture of the previous communicating artery on the right side. The brain angiography was performed. Subsequently, neurosurgery performs aneurysm clipper. It is admitted in critical care. The CT scan of control shows infarction plus severe cerebral edema on the right side, performing decompressive craniectomy (**Figure 16** (graph and images 1, 2 and 3)), without improvement, dying 8 days after admission.

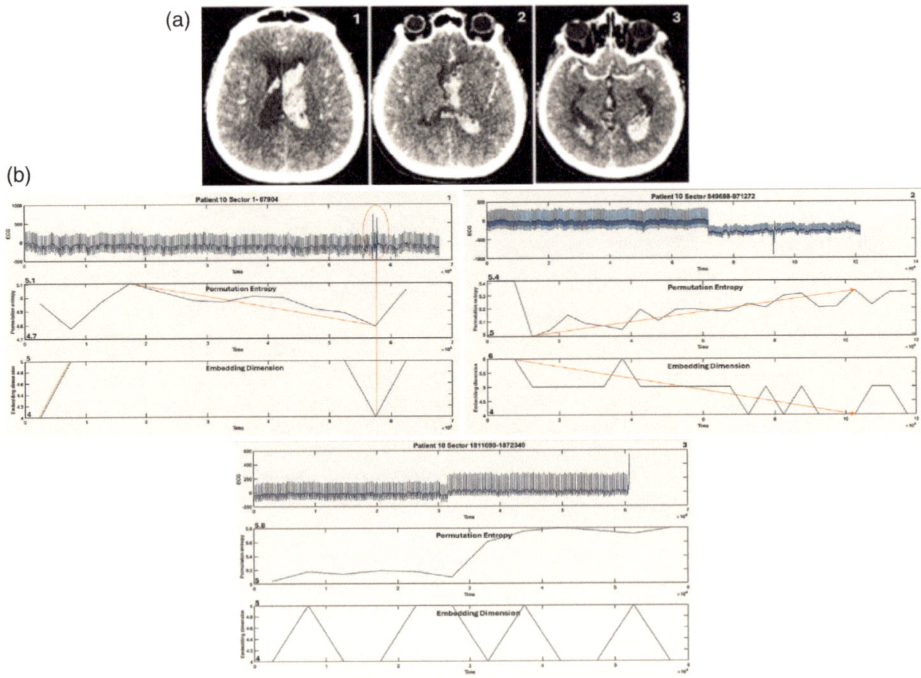

Figure 13.
(a) CT scan shows generalized severe cerebral edema and Fisher IV subarachnoid hemorrhage (images 1-3). (b) Graphics of permutation entropy and embedding dimension 1-3.

Unique graph. It is observed that the PE remains high, and paradoxically the ED also remains high at 7.

4.14 Patient 14

42-year-old female enters the hospital at the neurosurgery service, to perform a previously programmed endovascular procedure, as a treatment of an aneurysm in the previous communicating artery on the left side (**Figure 17** (graph and image 1)). During the procedure, the aneurysmatic sack is broken, with a leak of the contrast material to the subarachnoid space (images 2 and 3). It develops severe cerebral edema with quickly progressive neurological and systemic deterioration. On the sixth day, the brain death protocol begins [29], confirming this. The patient died 3 days later.

It is observed that the PE is altered by the disturbances shown in the ECG condensate, and that the ED is also modified by these disturbances, so it can be interpreted as a tonic behavior of this magnitude in 4.

4.15 Patient 15

Briefly described, not included in **Table 1** due to its peculiar behavior, but very significant in relation to its gravity and the result of its clinical evolution.

A 62-year-old male with cerebellar infarction of the left lobe (**Figure 18** (graphs and images 1 and 2)), with a score on the 4-point Glasgow scale, in ventilatory mechanical assistance. It was admitted to Critical Care in the immediate postoperative period (images 3 and 4). During its evolution, it was possible to remove mechanical

DOI: http://dx.doi.org/10.5772/intechopen.1012618

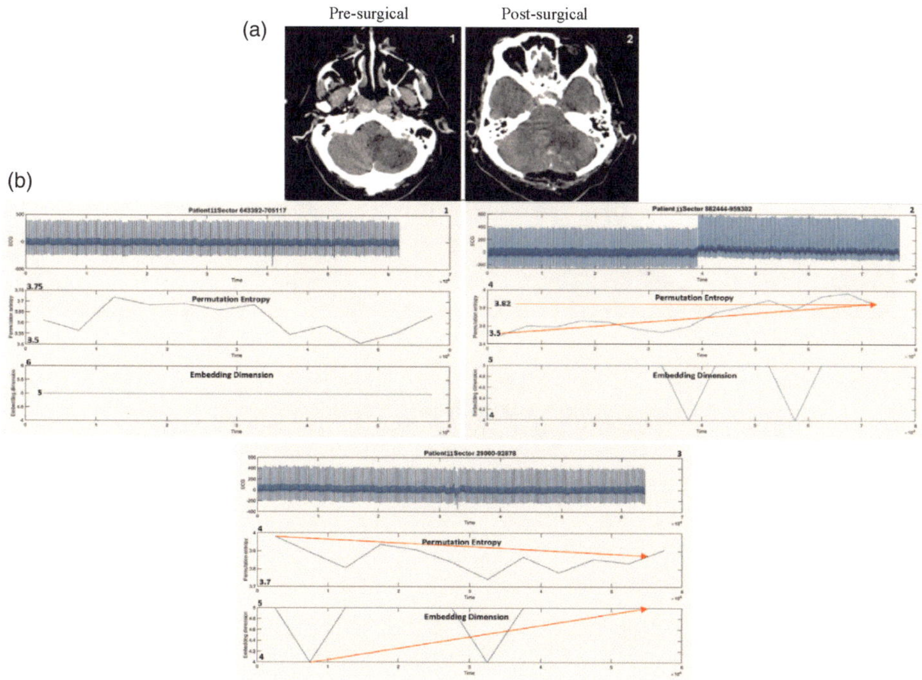

Figure 14.
(a) CT scan shows cerebellar infarction (image 1). Suboccipital craniectomy (image 2). (b) Graphics of permutation entropy and embedding dimension 1-3.

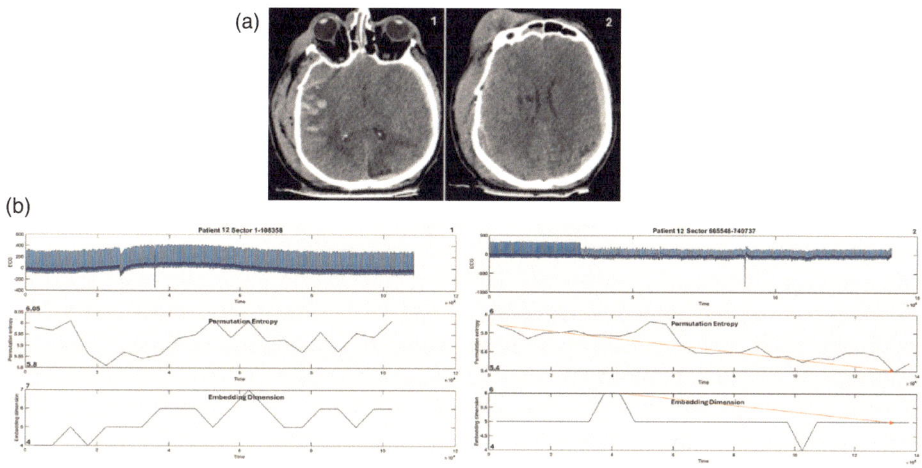

Figure 15.
(a) CT scan showed cerebral contusion and severe cerebral edema of the right side with suggestive areas of bleeding. (b) Graphics of permutation entropy and embedding dimension 1-2.

ventilation. From the neurological point of view, there was a very significant improvement with 13 points on the Glasgow scale. Therefore, his discharge of Critical Care was possible for improvement 10 days after his admission.

In graphs 1, 2 and 3, it can be seen that the ED tends to fluctuate between 4, 5 and 6. In graph 4, it can be seen how the PE decreases and the ED increases to 7 and remains.

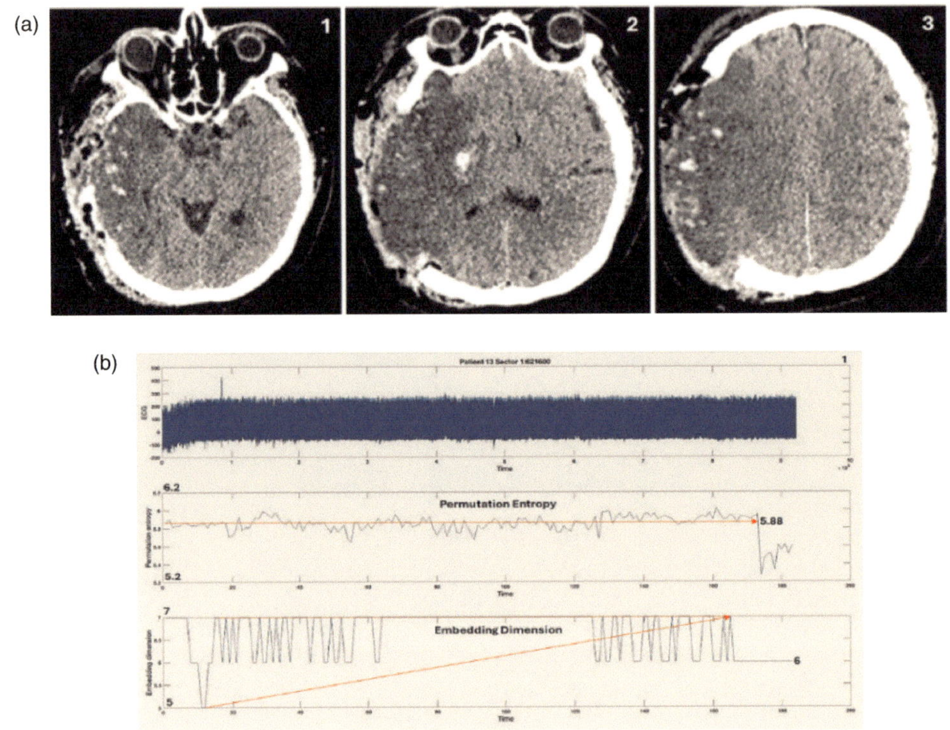

Figure 16.
(a) CT scan of control shows infarction plus severe cerebral edema, performing decompressive craniectomy (images 1-3). (b) Graph of permutation entropy and embedding dimension 1.

5. The following graphs correspond to time series of the 1-hour Holter recording (from 9 to 10 am), to 10 young subjects, sleeping in supine position

Briefly: Natural sleep is divided into two distinctive states, non-rapid eye movement (NREM) sleep, which is the period of parasympathetic predominance [33], and rapid eye movement (REM) sleep. NREM sleep is further divided into four stages, with stage 1 being the lightest and stage 4 or REM sleep the deepest; it is divided into two phases: tonic and phasic, with a regular pattern which is reached on average 90 minutes after NREM. Heart rate (HR) and systemic blood pressure (SBP) decrease during NREM and tonic REM sleep, while HR and SBP increase during phasic REM sleep [34]. In none of the 10 Holter recordings of the healthy subjects were periods of arrhythmia or tachycardia recorded.

The behavior of the magnitudes of PE and ED in healthy young humans oscillated within narrow ranges, PE from 5.1 to 5.27 and ED from 5.8 to 6.7. In subjects 2 and 6, an ED of up to 8 was recorded, while in the graphs of subjects 3 and 4, the behavior was tonic at ED of 6, so we consider that it could be related to NREM sleep. On the other hand, it is worth mentioning that in none of the healthy subjects there was an ED of 4 recorded. **Table 4**.

In the graphs of healthy men and females, PE ranged from 5.27 ± 0.4 to 5.1 ± 0.49, and ED from 6.7 ± 0.83 to 5.8 ± 0.6. In (**Figure 19**) graphs 3, 4, 6, and 7, ED showed a tonic behavior in 6. In graphs 2, 6 and 7, they showed an ED of up to 8.

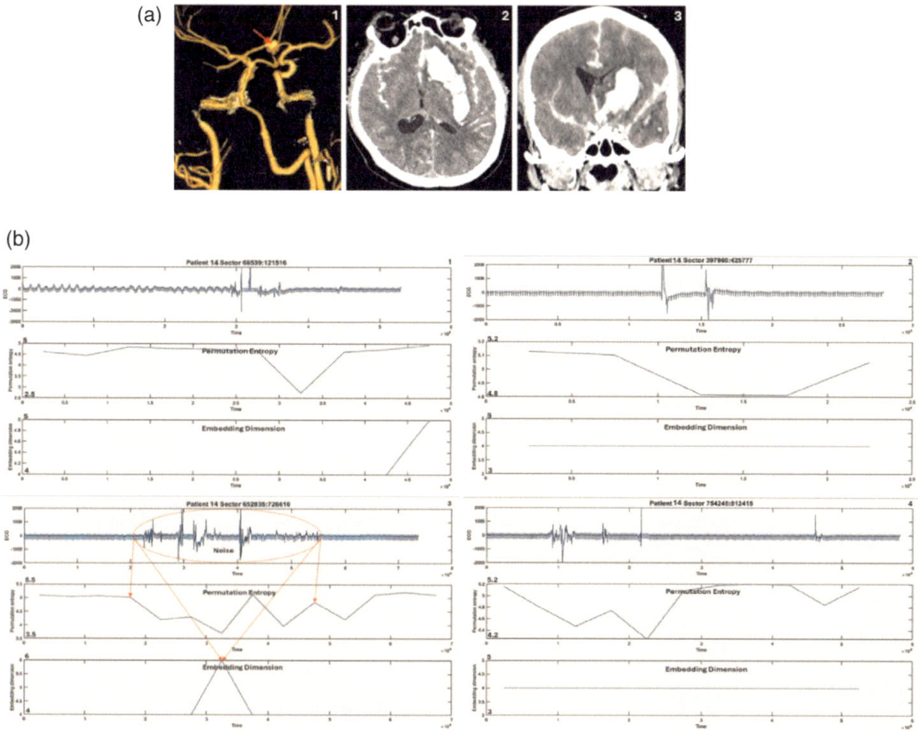

Figure 17.
(a) Aneurysm of the previous communicating artery on the left side (image 1). During the procedure, the aneurysmatic sack is broken, with a leak of the contrast material to the subarachnoid space (image 2 and 3). (b) Graphics of permutation entropy and embedding dimension 1-4.

6. Discussion

In this study, as in our previous work, we used an analog Holter monitor to analyze the dynamic behavior of the cardiac bioelectrical signal (CBES) in critically ill adult patients. To do so, we compared its Embedding Dimension (ED) with that obtained in a healthy 24-year-old man, as reported in Ref. [24]. With this prior information, we considered that, in critically ill patients as biological entities, the loss of information and the interaction of microsystems, subsystems, and systems could seriously compromise the dynamics of the CBES.

Entropy is considered fundamental to studying the content or loss of information or complexity of a system [11], so entropy, as a magnitude, is expected to increase as a result of the loss of information secondary to the severity of the patients. For these reasons, we measured this magnitude using Permutation Entropy (PE), comparing it with another magnitude, the Embedding Dimension (ED). The assumed behavior of both magnitudes in relation to the severity of the patients, with clinical evolution toward greater deterioration, was that the PE would increase and the ED would decrease, as shown in patient No. 1 (single graph), No. 2 graphs 3, No. 3 (single graph), No. 4 graphs 2 and 3, No. 8 graphs 6 and 9, and No. 10, graph 2, observing the bifurcation in the dynamic behavior, which can be interpreted as a state of greater severity. On the other hand, the opposite was also observed: As the ED increased, the PE decreased, as shown in patients No. 2 (graphs 1), No. 4 (graph 1), No. 8 (graphs 3, 5, 6, 8 and 9), No. 9 (graph 3) and No. 10 (graph 2). This can also be interpreted as an improvement in the outcome of some patients, when PE

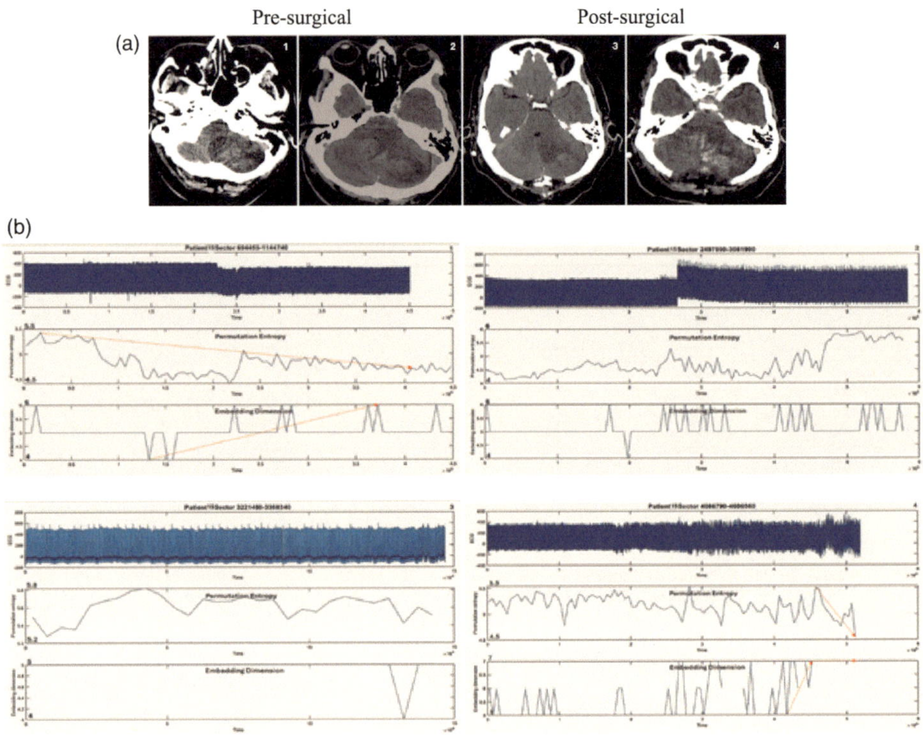

Figure 18.
(a) Cerebellar infarction of the left lobe (images 1 and 2). Decompressive craniectomy was performed (images 3 and 4). (b) Graphics of permutation entropy and embedding dimension 1-4.

Healthy young females and men					
	Permutation entropy			Embedding dimension	
N°	Magnitude	Magnitude	N°	Magnitude	Magnitude
1	4.7	4.95	1	6	7
2	5.55	4.8	2	6	8
3	4.95	5.2	3	6	6
4	5	5.2	4	6	6
5	5.65	5.85	5	5	6
6	4	4.5	6	7	8
7	5	5.6	7	6	7
8	5.35	5.5	8	5	6
9	5.6	5.68	9	6	7
10	5.2	5.5	10	5	6
	5.1 ± 0.49	5.27 ± 0.4		5.8 ± 0.6	6.7 ± 0.83

Table 4.
Age and sex of the 10 young subjects who were evaluated for the magnitudes of Permutation Entropy and Embedding Dimension.

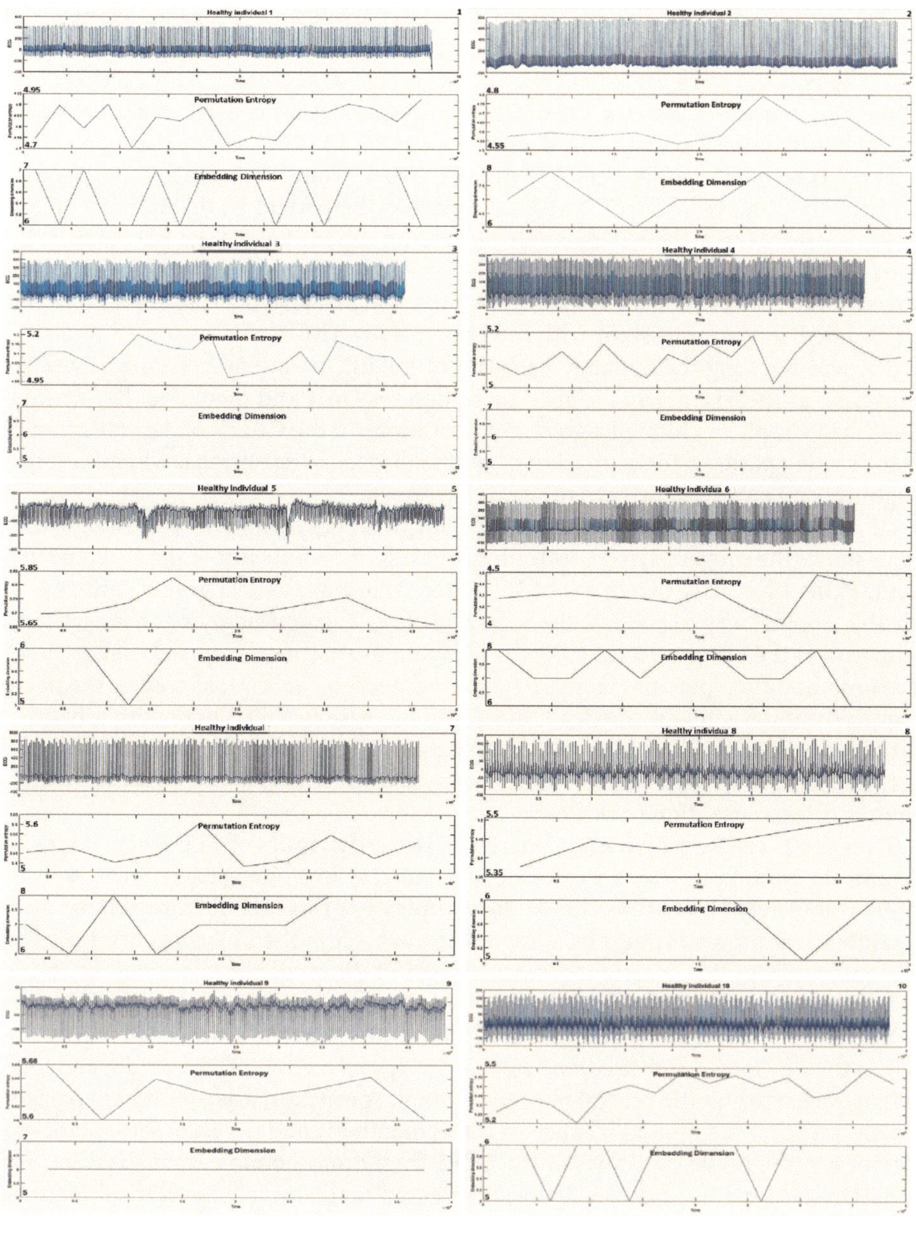

Figure 19.
Graphics of permutation entropy and embedding dimension 1-10.

decreases and ED increases, as in patients 2 (**Figure 1**), 4 (**Figure 1**), 8 (**Figures 3, 5, 8** and **9**) and 9 (**Figure 3**). This dynamic behavior of PE is not linear in all the cases analyzed; paradoxical behavior is also observed, as in patients 5 (**Figure 4**), 8 (**Figures 4** and **11**) (with an increase in both magnitudes) and 12 (**Figure 2**). Most of the paradoxical behavior showed that ED decreased and PE decreased as well. However, it is important to mention that, in the evolution of patients over time, only one hourly Holter recording is obtained every 24 hours, with a reporting interval of 23 hours for patients with a second Holter, the reporting interval is 46 hours and so on up to 16 Holter recordings in patient

No. 8. The graphs presented in the figures for each patient show a regular decrease in the ED value in all patients analyzed, and dynamic attempts to increase the ED are observed (e.g., graphs 1, 2, 3 and 4 for patient no. 6). This occurs in all patients over varying periods of time and frequency, with the ED decreasing to 4 in phasic behavior or even periods of tonic behavior also occurring at ED 4, as in patient no. 14.

This ED behavior ranges from 7 to 4, from 6 to 4, from 5 to 4, from 4 to 5, from 5 to 6, from 6 to 7, and from 5 to 7. The more severely the patient is, the more frequently the ED decreases to 4. This dynamic is repeated frequently until the ED reaches 4, when the patient dies. The short time that Patient 3 (single graph) was in critical care allowed us to obtain a 1-hour Holter recording, which showed a sustained PE of 5.7 and a tonic ED of 4, consistent with the patient's severity.

It is worth noting the dynamic behavior of the ED in Patient 2 (Graph 5) and 7 (Graph 2), as the ED briefly dropped to 3, then rose to 4 and then died. Both patients with ED in Graph 3 exhibited significant alterations in the CBES condensate, common when treating patients for aspiration of tracheobronchial secretions, which causes coughing, or also when the patient's body is moved.

Of the critical patients, Patient 1 stands out, with only 1 hour of recording, whose graph shows the progressive deterioration and loss of complexity of the CBES until her death. Patient No. 8, with a series of 16 Holter recordings and 11 graphs, offers a valuable opportunity to observe the behavior of both magnitudes over a longer period compared to the other patients studied, allowing to analyze the episodes of greatest deterioration of the patient (increase in PE and decrease in ED), as well as the periods of apparent improvement (decrease in PE and increase in ED). On the other hand, the particular behavior of both magnitudes in patient No. 14 with the documented brain death [29] draws attention to the linear behavior of ED in 4, in the 4 graphs, which requires greater attention to this fact.

Time series can help define and characterize the dynamics of a network, taking into account the context in which it is located, in our case in a biological context, so it may not be completely observable, due to the complexity of its architecture and its dynamic behavior over time, since it would be subject to the biological needs of the flow or loss of new information, generating and/or disappearing subnetworks to adapt to the needs of a new biological context [1], for example, health-disease or vice versa.

In patient N°14, the tonic behavior of the ED at 4 was sustained in all 4 graphs, as well as the PE (loss of information) strongly suggests a very important dependence, no longer exerted by the central nervous system, specifically the autonomic nervous system, and that the mutual balance of the sympathetic and parasympathetic nervous system, exerted in the dynamics of the CBES by the autonomic nervous system, stops modulating the cardiac biological work expressed as the initial bioelectrical sequence, and its mechanical biological consequence; the contraction, which allows the expulsion of blood from the cardiac cavities, to supply the demands to which a healthy or sick human being could be subjected.

The dynamic behavior of the PE and ED of patient N°15 opens the possibility of observing the improvement of a seriously ill individual, like patient N°11, by increasing the ED and maintaining it at 7, enabling rehabilitation.

7. Conclusion

The concept of biological signals has grown, becoming increasingly broader in its observations and content. The search in biological systems for their internal, structural and dynamic characteristics, obtaining relevant information from their components, interactions and observing them with other types of representative tools, has motivated the interest in understanding and explaining them. The cause or causes that can lead a human being to a serious state, as well as its course and outcome, have a very high unpredictability and heterogeneity, which seems to agree with the behavior of patients analyzed from the point of view of the decreasing the Embedding Dimension and increasing the Permutation Entropy. We assumed that the dynamics of all critically ill patients would behave in a very similar way. The analyzed dynamics show that this is not the case, since, based on this premise, the dynamics of both magnitudes were only observed in some graphs of the patients analyzed. If the concept of entropy implies the concept of information, it would be consistent that, when the cardiac bioelectrical signal decreases in embedding dimension during the patient's severity, this is reflected in an increase in permutation entropy (loss of information). Analysis of the behavior of the cardiac bioelectrical signal over time in critically ill patients reveals a mixture of dynamics, being both tonic and phasic in both magnitudes. The importance of understanding the heterogeneity of the complex dynamics of the cardiac bioelectric signal could allow us to identify the factors that influence its behavior, enabling discrete monitoring of the Embedding Dimension in patients to generate early warnings related to deterioration or improvement in their health status. In this regard, Holter recordings of more than one hour every 24 hours could be useful. The dynamic behavior of the bioelectric signal could be interpreted as an expression of the resilience and robustness of the system. Finally, it is worth mentioning that the Embedding Dimension allows us to obtain numerical values related to the deterioration or improvement of critically ill patients.

Author details

Pedro Eduardo Alvarado Rubio[1,2*] and Ricardo Lino Mansilla Corona[3,4]

1 Department of Cardiopulmonary, Critical Medicine and Intensive Therapy, National Autonomous University of Mexico (UNAM), Mexico

2 Department of Cardiopulmonary, Adult Intensive Therapy Unit of the Regional Hospital LIC. Adolfo López Mateos of the ISSSTE system in Mexico, Mexico

3 Peninsular Center for Humanities and Social Sciences, National Autonomous University of Mexico, Mexico

4 Center for Interdisciplinary Research in Sciences and the Humanities, National Autonomous University of Mexico, Mexico

*Address all correspondence to: pancreatitis2@gmail.com

References

[1] Bogdan P, Caetano-Anollés G, Jolles A, Kim H, Morris J, Murphy CA, et al. Biological networks across scales-the theoretical and empirical foundations for time-varying complex networks that connect structure and function across levels of biological organization. Integrative and Comparative Biology. 2022;**61**(6): 1991-2010. DOI: 10.1093/icb/icab069

[2] Kasum O, Perovic A, Jovanovic Á. Measures and metrics of biological signals. Frontiers in Physiology. 2018;**9**: 1707. DOI: 10.3389/fphys.2018.01707

[3] Wiener N. In: Hermann C, editor. Cybernetics or Control and Communication in the Animal and the Machine. Rue de la Sorbone, New York: The Technology Press Cambridge, Mass. John Wiley & Sons. Inc; 1948

[4] Césarman E, Brachfeld N, Estañol B. Entropy. The heart and the living systems. Ludus Vitalis. 2002;**10**(17–18):163

[5] Singh Y, N, Singh SK, Ray AK. Bioelectrical signals as emerging biometrics: Issues and challenges. International Scholarly Research Network. ISRN Signal Processing. 2012. Article ID 712032, 13 p. DOI: 10.5402/2012/712032

[6] Martinek R, Ladrova M, Sidikova M, Jaros R, Behbehani K, Kahankova R, et al. Advanced bioelectrical signal processing methods: Past, present and future approach—Part I: Cardiac signals. Sensors. 2021;**21**:5186. DOI: 10.3390/s21155186

[7] Cropper WH. Rudolf Clausius and the road to entropy. American Journal of Physics. 1986;**54**:1068. DOI: 10.1119/1.14740

[8] Willard Gibbs J. Elementary Principles in Statistical Mechanics Developed with Especial Reference to the Rational Foundation of Thermodynamics. New York: Charles Scribner's Sons, London: Edward Arnold, 1902. Statistical Physics Part II

[9] Prigogine I. Times, Structure and Fluctuations. Nobel conference-Lecture, 8 December 1977 By Université Libre de Bruxelles, Brussels, Belgium and the University of Texas at Austin, Austin, Texas, USA. Available from: https://www.nobelprize.org/uploads/2018/06/prigogine-lecture.pdf

[10] McMurry J. Organic Chemistry, 7 th ed. Cengage Learning Editores S.A. de C. V. 2008. ISBN-13: 978-607-481-349-4. ISBN-10: 607-481-349-32008

[11] Shannon C. A mathematical theory of communication. The Bell System Technical Journal. 1948;**27**:3

[12] Brissaud J-B. The meanings of entropy. Entropy. 2005;**7**(1):68-96. DOI: 10.3390/e7010068

[13] Sotolongo-Costa O, Rodriguez I. Entropy variation in a fractal phase space. Academia Letters. Article 662. 2021. DOI: 10.20935/AL662

[14] Captur G, Karperien AL, Hughes AD, Francis DP, Moon JC. The fractal heart - Embracing mathematics in the cardiology clinic. Nature Reviews Cardiology. 2017;**14**(1):56-64. DOI: 10.1038/nrcardio.2016.161. Epub 2016 Oct 6

[15] Takens F. Detecting strange attractors in turbulence. In: Rand D, Young LS, editors. Dynamical Systems and Turbulence, Warwick 1980, Lecture Notes in Mathematics. Vol. 898. Berlin, Heidelberg: Springer; 1981. DOI: 10.1007/BFb0091924

[16] Bandt C, Pompe B. Permutation entropy: A natural complexity measure for time series. Physical Review Letters. 2002;**88**(17):174102

[17] Tseng C-Y, Jack A. Tuszynski using entropy leads to a better understanding of biological systems. Entropy. 2010;**12**: 2450-2469. DOI: 10.3390/e12122450

[18] Parlitz U et al. Classifying cardiac biosignals using ordinal pattern statistics and symbolic dynamics. Computers in Biology and Medicine. 2012;**42**(3): 319-327

[19] Nicolaou N, Georgiou J. Detection of epileptic electroencephalogram based on permutation entropy and support vector machines. Expert Systems with Applications. 2012;**39**:202-209

[20] Cysarz D et al. Binary symbolic dynamics classifies heart rate variability patterns linked to autonomic modulations. Computers in Biology and Medicine. 2012;**42**:313-318

[21] Amigó J et al. Ordinal symbolic analysis and its application to biomedical recordings. Philosophical Transactions of the Royal Society A. 2015;**373**:20140091. DOI: 10.1098/rsta.2014.0091

[22] Small M. Applied Nonlinear Time Series Analysis: Applications in Physics, Physiology and Finance. Vol. 52. World Scientific; 2005

[23] Kantz H. A robust method to estimate the maximal Lyapunov exponent of a time-series. Physics Letters A. 1994;**185**:77-87. DOI: 10.1016/ 0375-9601(94)90991-1

[24] Rubio PEA, Corona RM, Vimbela LS, et al. Loss of complexity of the cardiac bioelectrical signal as an expression of patient outcomes. In: Interpreting Cardiac Electrograms - From Skin to Endocardium. London, UK: IntechOpen; 2017. DOI: 10.5772/intechopen.70144

[25] Grassbergert P, Procaccia I. Measuring the strangeness of strange attractors. Physica. 1983;**9D**:189-208

[26] Fraser A, Swinney H. Independent coordinates for strange attractors from mutual information. Physical Review A. 1986;**33**:pag. 1134

[27] Celluci C et al. Statistical validation of mutual information calculations: Comparison of alternative numerical methods. Physical Review E. 2005;**71**: 066208

[28] Albano AM, Muench J, Schwartz C, Mees AI, Rapp PE. Singular-value decomposition and the Grassberger-Procaccia algorithm. Physical Review A. 1988;**A38**:3017. DOI: 10.1103/ PhysRevA.38.3017

[29] Greer DM, Shemie SD, Lewis A, et al. Determination of brain death/death by neurologic criteria: The world brain death project. Journal of the American Medical Association. 2020;**324**(11):1078-1097. DOI: 10.1001/jama.2020.11586

[30] Fisher CM, Kistler JP, Davis JM. Relation of cerebral vasospasm to subarachnoid hemorrhage visualized by computerized tomographic scanning. Neurosurgery. 1980;**6**(1):1-9. DOI: 10.1227/00006123-198001000-00001

[31] Teasdale G, Maas AI, Lecky F, Manley G, Stocchetti N, Murray G, et al. The Glasgow coma scale at 40 years: Standing the test of time. Lancet Neurology. 2014;**13**(8):844-854. DOI: 10.1016/S1474-4422(14)70120-6

[32] Knaus WA, Draper EA, Wagner DP, Zimmermann JE. APACHE-II: A severity of disease classification system. Critical Care Medicine. 1985;**13**(10):819-829

[33] Kumar R. Cardiac autonomic regulation: Emerging biomarkers and clinical implications. In: Heart Rate Variability - Current Practices and Clinical Applications [Working Title]. London, UK: IntechOpen; 2025. DOI: 10.5772/intechopen.1012396

[34] Schupp M, Hanning CD. Physiology of sleep. British Journal of Anaesthesia | CEPD Reviews. 2003;**3**(3). DOI: 10.1093/bjacepd/mkg069

Chapter 3

Stability and Bifurcation in a Food Chain Model with the Allee Effect

Sarita Kumari and J. Omana

Abstract

This chapter describes the effects of specific intra-predator competition in a three-nutrient food chain model of a prey-dependent interacting population under Allee effect. We present a detailed mathematical analysis of the proposed food chain model to illustrate some ecological phenomenon and process under realistic environmental impact. In the context of your food chain model, the Allee effect may influence the behavior of prey, predator and super-predator populations, as their interactions are likely to be affected by changes in population density. We observe the biologically feasible equilibrium points of system and describe the behavior of system around each equilibrium point. Stability (local and global) and bifurcation (transcritical, saddle-node and Hopf-Andronov) analyses of the model are obtained. Global stability analysis is obtained through constructing particular Lyapunov functions. Numerical simulations show the relationship between prey, predator and super predator at different equilibrium points. In addition, numerical simulations with experimental parameter values confirm the analytical results and show the existence of bifurcation. Finally, a discussion of the analytical and numerical results is presented as the conclusion of this chapter.

Keywords: predator-prey, Allee effect, stability, bifurcation

1. Introduction

Food chain research is an important part of ecology and mathematical modeling of food chains has played a significant role in driving progress in modern ecology research. The trophic level of an organism is the position it holds in a food chain. In three-species (or tritrophic) food chains, the species at trophic level 1 is called prey, the species at trophic level 2 is called middle predator, and the species at trophic level 3 is called the top predator or super-predator. The Allee effect is named after U.S. behavioral scientist Warder Clyde Allee, and the term was coined by Odum. Since the late 1980s, the concept has grown in importance, but there remains a need for clear definitions and conceptual clarification. The Allee effect has numerous causes, including difficulty in finding mates, reduced anti-predator vigilance, problems with environmental conditioning and decreased defense against predators, among others.

The Allee effect is classified into two types based on the degree to which the per capita growth rate is reduced at low population densities: the strong Allee effect and the weak Allee effect. Unfortunately, theoretical ecology has largely remained silent on the impact of the Allee effect in tritrophic food chain models [1, 2]. This chapter focuses on equilibrium points, their stability, and bifurcation analysis for the model at each equilibrium point. First, we briefly introduce the construction of the model, its various equilibrium points, and their stability analysis. Then, based on this model, we perform numerical simulations.

There have been numerous modeling approaches related to species interaction in the literature, for example, see [3, 4]. The functional response term, which typically measures the amount of prey consumed by predators per unit time, is critical in developing a reliable mathematical model. Various functional responses, such as Holling type I to III [5–7], ratio-dependent [8–12], Beddington-DeAngelis [13–15], and Crowley-Martin function response [16, 17], have been developed and used. In population dynamics modeling, two types of models are commonly used: discrete [18, 19] and continuous-time models [15, 17]. Food chain models are aware of the interaction between two or more prey and predators [20–23], and contain a variety of food chain models with varying interactions and the associated qualitative analysis. In this regard, a common model is a top predator predating on an intermediate predator, who then predates on a prey species. Upadhyay and Naji [24] studied the local and global stability effects and considered such a model. Ref. [21] investigated the effect of mutual interference among second-level predators *via* Beddington-DeAngelis functional response, where the considered model exhibited chaotic behaviors by varying suitable parameters. Recently, researchers discussed some important results concerning the study of a dynamical system in various fields, which were published in [25–28].

In this study, we have investigated a cubic two-dimensional coupled logistic equation in a discrete framework. This model captures the nonlinear interactions between two species while incorporating cubic growth terms, which allow for a richer set of dynamical behaviors compared to the standard logistic model. Our analysis focuses on the system's fixed points, their stability, and the emergence of complex dynamics such as bifurcations and chaos under varying parameter regimes [29]. The chapter "logistic models for symbiosis, predator-prey and competition" [30] explores how logistic equations can model various ecological interactions. It demonstrates the role of coupled nonlinear systems in analyzing species coexistence and stability. This work provides a valuable foundation for studying complex biological dynamics using logistic frameworks [30]. Gilani et al. [31] investigated the influence of Allee effects and diffusion-driven instability on predator-prey interactions. Their study provides valuable insights into the conditions that promote ecological persistence and those that may lead to extinction, highlighting the complex dynamics introduced by spatial and population density factors.

In the current chapter, we focus on the local and global stability of equilibria, the persistence of the three species, and the bifurcations of the system. Note that the initial model and the nondimensionalized model share the same number of variables-prey density, predator density, and super-predator density but differ in the number of parameters, as the nondimensional model is a simplified version. The rest of the chapter is organized as follows: Section 3 presents some preliminary results. Section 4 discusses the behavior of the system around all positive equilibrium points, including local stability, global stability, and bifurcations. Sections 5 and 6 presents the detailed derivations and illustrates the analytical findings through numerical simulations. Finally, Section 7 provides a concluding discussion and interpretation of our analytical and numerical results.

2. Mathematical model formulation

A food chain model with Allee effect on prey can be written using the following system of ordinary differential equations (ODEs):

$$\begin{cases} \dfrac{dx}{dt} = \dfrac{rx}{x+n} + \left(1 - \dfrac{x}{k}\right)(x-a) - m_1xy, \\ \dfrac{dy}{dt} = e_1m_1xy - m_2yz - d_1y, \\ \dfrac{dz}{dt} = e_2m_2yz - d_2z, \end{cases} \tag{1}$$

$$x(0) > 0 \ , \ y(0) > 0, z(0) > 0. \tag{2}$$

where $x(t), y(t)$ and $z(t)$ stand for the prey density, predator density, and super-predator density, respectively, at time t; and r, k, m_i, e_i, d_i, a and n are positive constants that stand for prey intrinsic growth rate, carrying capacity of the environment, consumption rates, conversion rates, predator death rates, a constant and Allee constant, respectively.

The system of Eq. (1) has been non-dimensionalized according to $u = \frac{x}{k}, \ v = \frac{y}{ke_1}, \ w = \frac{z}{ke_1e_2}, \ \tau = rt$ which lead them to

$$\begin{cases} \dfrac{du}{d\tau} = \dfrac{u}{ku+n}(1-u)(ku-a) - M1uv, \\ \dfrac{dv}{d\tau} = M_1uv - M_2vw - P_1v, \\ \dfrac{dw}{d\tau} = M_2vw - P_2w, \end{cases} \tag{3}$$

$$u(0) > 0 \ , \ v(0) > 0, w(0) > 0, \tag{4}$$

where the dimensionless parameters are given by

$$M_1 = \frac{e_1m_1k}{r} \ , \ M_2 = \frac{m_2e_1e_2k}{r}, \quad P_1 = \frac{d_1}{r} \quad , P_2 = \frac{d_2}{r}. \tag{5}$$

3. Mathematical analysis of model system

3.1 Equilibria

The system (1) has the following positive equilibria:

- Trivial equilibrium: $C_1(0,0,0)$.
- Axial equilibrium: $C_2(1,0,0)$ and $C_3\left(\frac{k}{a},0,0\right)$.
- Boundary equilibrium: $C_4(u_1, v_1, 0)$.

  where $u_1 = \frac{p_1}{M_1}, v_1 = \frac{(1-u_1)(ku_1-1)}{(ku_1+n)M_1}$

- Positive interior equilibrium: $C_5(u, v, w)$

  where $v = \frac{P_2}{M_2}$, $M_1 v = \frac{(1-u)(ku-a)}{(ku+n)}$ (*).

(*) leads to the equation: $-ku^2 + \left(k + a - \frac{P_2 M_1 k}{M_2}\right)u - \left(a + \frac{nP_2 M_1}{M_2}\right) = 0$

$$\Delta = \left(k + a - \frac{P_2 M_1 k}{M_2}\right)^2 - 4k\left(a + \frac{nP_2 M_1}{M_2}\right).$$

Case (a) If $\Delta > 0$: Two points $\left(u_2, \frac{P_2}{M_2}, w_2\right)$, $\left(u_3, \frac{P_2}{M_2}, w_3\right)$ where $u_2 = \frac{k+a-\frac{P_2 M_1 k}{M_2}}{2} + \frac{\sqrt{\Delta}}{2}$, $u_3 = \frac{k+a-\frac{P_2 M_1 k}{M_2}}{2} - \frac{\sqrt{\Delta}}{2}$, $w_2 = \frac{M_1 \; u_2+P_1}{M_2}$, $w_3 = \frac{M_1 \; u_3+P_1}{M_2}$.

Case (b) If $\Delta < 0$: the equilibrium point cannot exist.

Case (c) If $\Delta = 0$: only one point $\left(u_4, \frac{P_2}{M_2}, w_4\right)$, where $u_4 = \frac{k+a-\frac{P_2 M_1 k}{M_2}}{2}$, $w_4 = \frac{M_1 \; u_4+P_1}{M_2}$.

Boundedness: The solution $s(\tau) = u + v + w$ of system which initiate in R^3_+ (that is, $s_0 \in R^3_+$) is uniformly bounded.

4. Stability and bifurcation analysis

This section discusses the local stability, global stability, and bifurcation analysis of the system. The detailed derivation process and proof contain a detailed description of the proofs for the stability and bifurcation results. Only the results and conclusion are reported here.

4.1 Behavior of the model system around the trivial $C_1(0,0,0)$

Lemma 4.1.

i. Local stability: We obtain that three eigenvalues of systems around $C_1(0,0,0)$ are $-a/n$, $-P_1$, $-P_2$ through nondimensionalizing. The all three eigenvalues are negative real root because all parameters are positive constants and therefore the hyperbolic equilibrium point $C_1(0,0,0)$ is locally asymptotically stable at any condition.

ii. Hopf and Saddle-node bifurcation: It does not have saddle-node, Hopf bifurcation since eigenvalues are all real numbers.

4.2 Behavior of the model system around $C_2(1,0,0)$

The Jacobian matrix of Eq. (3) at $C_2(1,0,0)$:

$$J = \begin{pmatrix} J_{11} & J_{12} & J_{13} \\ J_{21} & J_{22} & J_{23} \\ J_{31} & J_{32} & J_{33} \end{pmatrix},$$

where

$$\begin{cases} J_{11} = \dfrac{a-k}{k+n}, \ J_{12} = -M_1, \ J_{13} = 0, \\ J_{21} = 0, \ J_{22} = M_1 - P_1, \ J_{23} = 0, \\ J_{31} = 0, \ J_{32} = 0, \ J_{33} = -P_2. \end{cases} \tag{6}$$

Lemma 4.2.

i. Local stability: We obtain the three eigenvalues of systems around $C_2(1{,}0{,}0)$ are $(a-k)/(k+n), M_1 - P_1, -P_2$ through nondimensionalizing. All three eigenvalues are negative real roots, and the hyperbolic equilibrium point $C_2(1{,}0{,}0)$ is locally asymptotically stable if and only if $a < k, M_1 < P_1$.

ii. Hopf and Saddle-node Bifurcation: It does not have saddle-node, Hopf-bifurcation since eigenvalues are all real numbers.

iii. Transcritical Bifurcation: This system cannot experience transcritical bifurcation since $W^T \frac{\partial f}{\partial M_1}\left(\vec{x}_0, M_1\right) = 0$ and $W^T\left[D\frac{\partial f}{\partial M_1}\left(\vec{x}_0, M_1\right)\vec{v}\right] \neq 0$ but $W^T\left[D^2 f\left(\vec{x}_0, M_1\right)(v,v)\right] = 0$.

iv. Global stability condition: $M_1 < P_1$, so the local stability of the system ensures its global stability around the interior equilibrium point C_2.

4.3 Behavior of the model system around $C_3\left(\frac{a}{k}, 0, 0\right)$

The Jacobian matrix of Eq. (3) at $C_3\left(\frac{a}{k}, 0, 0\right)$:

$$\begin{aligned} J_{11} &= \frac{a(k-a)}{k(a+n)}, \ J_{12} = -M_1 * \frac{a}{k}, \ J_{13} = 0, \\ J_{21} &= 0, \ J_{22} = M_1 * \frac{a}{k} - P_1, \ J_{23} = 0, \\ J_{31} &= 0, J_{32} = 0, J_{33} = -P_2. \end{aligned} \tag{7}$$

Lemma 4.3.

i. Local stability: We obtain the three eigenvalues of systems around $C_3\left(\frac{a}{k}, 0, 0\right)$ are $\frac{a(k-a)}{k(a+n)}, \frac{aM_1}{k} - P_1, -P_2$ through nondimensionalizing. All three eigenvalues are negative real roots, and the hyperbolic equilibrium point $C_3\left(\frac{a}{k}, 0, 0\right)$ is locally asymptotically stable if and only if $a > k, \ aM_1 < kP_1$.

ii. Hopf-bifurcation and saddle-node: It does not have saddle node, Hopf bifurcation contains eigenvalues that are all real numbers.

iii. Transcritical bifurcation: This system cannot experience transcritical bifurcation since

Case 1 ($a = k$): $W^T \frac{\partial f}{\partial M_1}\left(\vec{x}_0, M_1\right) = 0$ and $W^T\left[D\frac{\partial f}{\partial M_1}\left(\vec{x}_0, M_1\right)\vec{v}\right] \neq 0$ but $W^T\left[D^2 f\left(\vec{x}_0, M_1\right)(v,v)\right] = 0$.

Case 2 $(M_1 = kP_1/a)$: $W^T\left[D\frac{\partial f}{\partial M_1}\left(\vec{x}_0, M_1\right)\vec{v}\right] = 0.$

iv. Global stability condition: $a > k,\ \ aM_1 < kP_1$, so local stability of the system ensures its global stability around the interior equilibrium point C_3.

4.4 Behavior of the model system around $C_4(u_1, v_1, 0)$

The Jacobian matrix of Eq. (3) at $C_4(u_1, v_1, 0)$:

$$\begin{gathered} J_{11} = m_{11},\ J_{12} = -M_1u_1,\ J_{13} = 0, \\ J_{21} = M_1v_1,\ J_{22} = \mu_1u_1 - P_1,\ J_{23} = M_1v_1, \\ J_{31} = 0,\ J_{32} = 0,\ J_{33} = M_2v_1 - P_2, \end{gathered} \tag{8}$$

Lemma 4.4.

i. Local stability: The equilibrium point $C_4(u_1, v_1, 0)$ is locally asymptotically stable if the conditions $1 + \frac{a}{k} - 2\frac{P_1}{M_1} - \frac{\left(1-\frac{P_1}{M_1}\right)\left(\frac{P_1}{M_1}k-a\right)}{\frac{P_1}{M_1}k+n} < 0$ and $v_1 < \frac{P_2}{M_2}$, using Routh-Hurwitz theorem.

ii. Saddle-node bifurcation: The system experiences a saddle-node bifurcation around $M_1 = M_1^{[sn_1]}$ where $\left(M_1^{[sn_1]} - P_1\right)\left(P_1k - aM_1^{[sn_1]}\right) - P_2\left(P_1k + nM_1^{[sn_1]}\right) = 0.$

iii. Hopf-bifurcation: The system enters into Hopf-bifurcation around $M_1 = M_1^{[hb_1]}$ where $M_1^{[hb_1]}$ satisfy the equations:

$$1 + \frac{a}{k} - 2\frac{P_1}{M_1^{[hb_1]}} - \frac{\left(P_1 - M_1^{[hb_1]}\right)\left(P_1k - aM_1^{[hb_1]}\right)}{\left(P_1k + nM_1^{[hb_1]}\right)M_1^{[hb_1]}} = 0$$

and

$$\frac{u_1}{ku_1+n}\left[\frac{u_1}{a}(ku_1+n)^3(n+a)(k+n) - (u_1k+n)^2\right]^2 - 4M_1(1-u_1)(u_1k-a) < 0.$$

4.5 Behavior of the model system around C_5

$$J_{11} = m_{11}, J_{12} = -M_1u, J_{13} = 0,\ J_{21} = P_2\frac{M_1}{M_2}, J_{22} = 0, J_{23} = -P_2,\ J_{31} = 0, J_{32} = M_1u - P_1, J_{33} = 0.$$

Lemma 4.5.

i. Local stability: The equilibrium point $C_5(u, v, w)$ is locally asymptotically stable if satisfy the conditions $\frac{P_1}{M_1} < u < \frac{P_1}{M_1 + m_{11}\frac{M_1^2}{M_2}}$ and $m_{11} < 0$ where

$$m_{11} = \frac{\left[\frac{P_2 M_1 k}{M_2} - (k+a)\right] u + 2\left(a + \frac{n P_2 M_1}{M_2}\right)}{uk+n}.$$

Note: u equals to u_2, u_3 and u_4 according to case (a) and case (c), respectively, in four Preliminary results.

ii. Hopf-bifurcation: this system experiences Hopf-bifurcation around $M_1 = M_1^{[hb_2]}$ where $M_1^{[hb_2]}$ satisfy the equation $\left(\frac{m_{11}\left(M_1^{[hb_2]}\right)^2}{M_2} + M_1^{[hb_2]}\right) u - P_1 = 0.$

5. Detailed derivation process and proof

5.1 Proof of lemma 4.2

Proof: $C_2 \ (1{,}0{,}0)$ global stability condition is:
when $(k_1 M_1 - k_2 P_1) < 0$ and $u > 0 (a \ll k)$

$$\begin{aligned}
L_1 &= k_1\left(u - u^* - u^* \, ln \, \frac{u}{u^*}\right) + k_2 v + k_3 w, (k_1, k_2, k_3 > 0), \\
\frac{dL_1}{dt} &= k_1\left(1 - \frac{u^*}{u}\right)\frac{du}{dt} + \ + k_2\frac{dv}{dt} + k_3\frac{dw}{dt} \quad and \quad u = 1, v = 0, w = 0, \\
\frac{dL_1}{dt} &= k_1(u-1)\left[\frac{(1-u)(uk-a)}{uk+n} - M_1 v\right] \\
&+ k_2 v(M_1 u - M_2 w - P_1) + k_3(M_2 v - P_2)w, \\
&= -k_1(u-1)^2\frac{(uk-a)}{uk+n} - k_1(u-1)M_1 v \\
&+ k_2 M_1 uv - k2M_2 wv - k2P_1 v + k3M_2 vw - k3P_2 w, \\
&= -k_1(u-1)^2\frac{(uk-a)}{uk+n} + k_1 M_1 v - k_1 M_1 uv + k_2 M_1 uv \\
&- k_2 M_2 wv - k_2 P_1 v + k_3 M_2 vw - k_3 P_2 w.
\end{aligned} \tag{9}$$

Let us choose $k_1 = k_2 = k_3 > 0$, then

$$\frac{dL_1}{dt} = -k_1(u-1)^2\left(\frac{uk-a}{uk+n}\right) + (k_1 M_1 - k_2 P_1)v - k_3 P_2 w < 0. \tag{10}$$

Also $\frac{dL_1}{dt} = 0 \ when \ \ C_2 \ (1{,}0{,}0)$. Hence, according to the Lyapunov-Lasalle invariance principle, $C_2 \ (1{,}0{,}0)$ is globally asymptotically stable.

5.2 Proof of lemma 4.3

Proof: $C_3\left(\frac{a}{k}{,}0{,}0\right)$ global stability condition is:
when $\frac{a}{k} M_1 < P_1 \ and \ a > k,$

$$\begin{aligned}
&L_2 = k_1\left(u - u^* - u^* \ln \frac{u}{u^*}\right) + k_2 v + k_3 w \quad (k_1, k_2, k_3 > 0),\\
&\frac{dL_2}{dt} = k_1\left(1 - \frac{u^*}{u}\right)\frac{du}{dt} + \ +k_2\frac{dv}{dt} + k_3\frac{dw}{dt} \quad and \ u = \frac{a}{k}, v = 0, w = 0,\\
&\frac{dL_2}{dt} = k_1\left(1 - \frac{a}{ku}\right)\left[\frac{(1-u)(uk-a)}{uk+n} - M_1 v\right] + k_2 v(M_1 u - M_2 w - P_1) + k_3(M_2 v - P_2)w\\
&= k_1(u-a)^2\frac{(1-u)}{(uk+n)k} - k_1 M_1 v\frac{(uk-a)}{k} + k_2 M_1 uv - k_2 M_2 wv - k_2 P_1 v + k_3 M_2 vw - k_3 P_2 w,
\end{aligned} \tag{11}$$

Let us choose $k_1 = k_2 = k_3 > 0$ and $u > 1$.

Then $\frac{dL_2}{dt} = k_1(u-a)^2\frac{(1-u)}{(uk+n)k} + \left[\frac{a}{k}k_1 M_1 - k_2 P_1\right]v - k_3 P_2 w < 0$.

Only satisfied when $\frac{a}{k}M_1 < P_1$ *and* $a > k$.

Also $\frac{dL_2}{dt} = 0$ *when* $C_3\left(\frac{a}{k},0,0\right)$. Hence by Lyapunov-Lasalle's invariance principle, $C_3\left(\frac{a}{k},0,0\right)$ is globally asymptotically stable.

5.3 Proof of lemma 4.4

Proof: $C_4(u_1, v_1, 0)$

As our Jacobian matrix is

$$\begin{aligned}
J_{11} = m_{11},\ J_{12} = -M_1 u_1,\ J_{13} = 0,\\
J_{21} = M_1 v_1,\ J_{22} = \mu_1 u_1 - P_1,\ J_{23} = M_1 v_1,\\
J_{31} = 0,\ J_{32} = 0,\ J_{33} = M_2 v_1 - P_2,\\
|J_{C4} - \lambda I| = 0.
\end{aligned}$$

Follow the Routh-Hurwitz theorem, when $J_{11} < 0$ *and* $J_{33} < 0$, the equilibrium point $C_4(u_1, v_1, 0)$ is locally asymptotically stable. So the condition is $m_{11} < 0, v_1 < \frac{P_2}{M_2}$,

where $m_{11} = \frac{\frac{P_1}{M_1}k}{\frac{P_1}{M_1}k+n}\left[1 + \frac{a}{k} - 2\frac{P_1}{M_1} - \frac{\left(1-\frac{P_1}{M_1}\right)\left(\frac{P_1}{M_1}k-a\right)}{\frac{P_1}{M_1}k+n}\right]$.

For the Hopf-bifurcation, we have the characteristic equation

$\left[\frac{\left(1-\frac{P_1}{M_1}\right)\left(\frac{P_1}{M_1}k-a\right)}{\frac{P_1}{M_1}k+n}\frac{M_2}{M_1} - P_2 - \lambda\right]\left[(m_{11}-\lambda)(-\lambda) + P_1\frac{\left(1-\frac{P_1}{M_1}\right)\left(\frac{P_1}{M_1}k-a\right)}{\frac{P_1}{M_1}k+n}\right] = 0$ since $|J_{C4} - \lambda I| = 0$.

We need to have complex root of eigenvalue, then $\Delta < 0$, which means

$$\frac{u}{ku+n} - \left(\frac{u}{(ku+n)^3(n+a)}(n+a)(k+n) - (uk+n)^2\right)^2 - 4M_1(1-u)(uk-a) < 0.$$

when $u = \frac{P_1}{M_1}$.

Besides, when Hopf-bifurcation occurs, $J_{11} = 0$ to ensure the purely imaginary roots which means $m_{11} = 0$.

Then $1 + \frac{a}{k} - 2\frac{P_1}{M_1} - \frac{\left(1-\frac{P_1}{M_1}\right)\left(\frac{P_1}{M_1}k-a\right)}{\frac{P_1}{M_1}k+n} = 0$ after simplifying it by linear algebra, when $M_1 = M_1^{[hb_1]}$.

So the second condition is $(an + kn + ak)\ {M_1}^2 + \left(2k^2P_1 - 2P_1kn\right)\ M_1 - {P_1}^2k^2 = 0$, where $M_1 = M_1^{[hb_1]}$ in this equation.

Next we need to show $Re\left(\frac{d\lambda}{dM_1}\right) = \frac{\frac{1}{2}\partial J_{11}}{\partial M_1} \neq 0$,

which satisfied when $\frac{\partial J_{11}}{\partial M_1} =$

$$\left(nP_1k(P_1k + nM_1)^2\left(1 + \frac{a}{k} - \frac{2P_1}{M_1} - \frac{\left(1 - \frac{P_1}{M_1}\right)\left(\frac{P_1k}{M_1} - a\right)}{\frac{P_1k}{M_1} + n}\right) + \left(\frac{\frac{P_1k}{M_1}}{\frac{P_1k}{M1} + n}\right)\right)$$

$$\frac{P_1}{\left(M_1(P_1k + nM_1)^2\right)}\left[(2n + a) - k(a + n)M_1^2 + 2P_1knM_1 + P_1^2k^2\right].$$

Since $M_1 = {M_1}^{[hb_1]}$, $\left[1 + \frac{a}{k} - 2\frac{P_1}{M_1} - \frac{\left(1 - \frac{P_1}{M_1}\right)\left(\frac{P_1}{M_1}k - a\right)}{\frac{P_1}{M_1}k + n}\right] = 0.$

Then $\frac{\partial J_{11}}{\partial M_1} = \left(\frac{\frac{P_1}{M_1}k}{\frac{P_1}{M_1}k + n}\right) * \left(\frac{P_1}{M_1(P_1k + nM_1)^2}\right)\left[(2n + a) - k(a + n){M_1}^2 + 2P_1knM_1 + {P_1}^2k^2\right].$

$Re\left(\frac{d\lambda}{dM_1}\right) \neq 0$ when $M_1 = M_1^{[hb1]}$ since $\left[1 + \frac{a}{k} - 2\frac{P_1}{M_1} - \frac{\left(1 - \frac{P_1}{M_1}\right)\left(\frac{P_1}{M_1}k - a\right)}{\frac{P_1}{M_1}k + n}\right] = 0$ and $\left[(2n + a) - k(a + n){M_1}^2 + 2P_1knM_1 + {P_1}^2k^2\right]$ cannot be zero at the same time.

Note:

$$1 + \frac{a}{k} - \frac{2P_1}{M_1} - \frac{\left(1 - \frac{P_1}{M_1}\right)\left(\frac{P_1k}{M_1} - a\right)}{\frac{P_1k}{M_1} + n} = 0$$

is equivalent to the equation

$$(an + kn + ak)^2 + \left(2k^2P_1 - 2P_1kn\right)M_1 - {P_1}^2k^2 = 0.$$

Then the system exhibits a Hopf-bifurcation in the first octant leading to a family of solutions that bifurcates from $C_4(u_1, v_1, 0)$ for suitable values of M_1 in a neighborhood of $M_1 = M_1^{[hb1]}$.

5.4 Proof of lemma 4.5

Proof: $C_5\left(u, \frac{P_2}{M_2}, \frac{M_1u - P_2}{M_2}\right)$

for u_2, u_3, u_4, the form of cubic equation about λ is same: $\lambda^3 + A_1\lambda^2 + A_2\lambda + A_3 = 0$, (6)

where $A_1 = -m_{11}, A_2 = \frac{P_2}{M_1^2}u, A_3 = P_2(M_1u - P_1)$.

By Routh-Hurwitz criterion, a set of necessary and sufficient conditions for characteristic equation to have Hopf-bifurcation is $A_2 > 0$ and $A_1A_2 - A_3 > 0$. $A_2 > 0$ is clearly the case since the roots of the characteristic equation are larger than 0. Then, $A_1A_2 - A_3 > 0$ is equivalent to the fact that the characteristic Eq. (6) has two purely imaginary roots. Now, at $M_1 = M_1^{[hb_2]}$, we have $(\lambda^2 + A_2)(\lambda + A_1) = 0$ which has three roots $\lambda_1 = i\sqrt{A_2}, \lambda_2 = -i\sqrt{A_2}, \lambda_3 = -A_1$. We verify the transversality condition $Re\left(\frac{d\lambda_i}{dM_1}\right) \neq 0$ when $M_1 = M_1^{[hb_2]}$ $(i = 1, 2)$ by numerical method.

6. Numerical simulations

6.1 Parameter value range

The parameter values of the eight parameters $r,\ k, m_1, m_2, e_1, e_2, d_1$ and d_2, which are all positive constants are shown in **Table 1** below. Specially, there are two positive constants a and n, which related to Allee effect. Both would be much smaller than k ideally.

To deal with the local stability and the bifurcation, we analyzed these figures numerically and plotted them using MATLAB and MatCont of system. During the next section of our report, the specific values we defined are:

$$\begin{gathered} m_1 = 0.09, 0.2, 0.02, 0.016; m_2 = 0.15, 0.2; \\ e_1 = 0.03, 0.02, 0.16, 0.05; e_2 = 0.45, 0.9, \\ 0.1, 0.08; d_1 = 0.16, 0.28, 0.08; d_2 = 0.2, \\ 0.04, 0.16; r = 0.02, 0.04, 0.8; k = 50, 100, 200. \end{gathered} \tag{12}$$

Given the nondimentionalizing transaction equations corresponding to those parameters above, we can get the value range of the corresponding nondimentionalized parameters $(M_1, M_2, P_1,\ P_2)$ in this system, which are shown in **Table 2**. According to the range we have taken, eight main parameters are considered before nondimentionalizing, which are shown in **Table 1**. We took limit values and applied them in derivation calculations; the nondimentionalized parameter values used in our next study are within the corresponding ranges given in **Table 2**. In the results of our report, we defined the specific values below:
$M_1 = 6.75, 5, 0.2; M_2 = 5.06, 4.5, 0.2; P_1 = 8, 7, 0.1; P_2 = 10, 1, 0.2$.

Corresponding meaning	Parameter	Minimum	Maximum
Prey intrinsic growth rate	r	0.02	0.8
Carrying capacity of the environment	k	50	500
Consumption rate of prey	m_1	0.01	0.2
Consumption rate of predator	m_2	0.01	0.2
Conversion rate of predator	e_1	0.01	1
Conversion rate of super-predator	e_2	0.01	1
Death rate of predator	d_1	0.01	0.2
Death rate of super-predator	d_2	0.01	0.2

Table 1.
Biological meanings of parameters used in the model system (1).

Parameter	Corresponding formulas	Minimum	Maximum
M_1	$\frac{m_1 e_1 k}{r}$	6.25×10^{-3}	5×10^3
M_2	$\frac{m_2 e_1 e_2 k}{r}$	6.25×10^{-5}	5×10^3
P_1	$\frac{d_1}{r}$	1.25×10^{-3}	10
P_2	$\frac{d_2}{r}$	1.25×10^{-3}	10

Table 2.
The range of three parameters of the model system (1) after nondimensionalizing.

6.2 Stability and bifurcation

In this section, defining the reasonable value of the parameters (both dimentionalized and nondimentionalized) and substituting them into the model to validate the various conditions (about fitting the stability and bifurcation of different points, which we have analyzed before this section) and reflecting the important characteristics of the system more significantly based on realistic observations is one of our main purposes. Next step, we conducted specific stability studies results based on five different types of points: $C_1(0,0,0), C_2(1,0,0), C_3\left(\frac{a}{k},0,0\right)$, $C_4\left(\frac{P_1}{M_1}, \frac{f\left(\frac{P_1}{M_1}\right)}{M_1}, 0\right)$ and $C_{5,6}(u,v,w)$ (note: u, v, w are formulas constructed by the parameters), as shown in **Tables 1** and **2**. The details are shown before type 5. Then we plotted our model to find the relationship between three species (prey, predator, and super-predator) trends based on time (nondimentionalized) under different conditions and did numerical analysis.

6.2.1 Type 1 $C_1(0,0,0)$

Figure 1 shows the global stability of the nondimentionalized system and the relationship between the three species based on time t at the point $C_1(0,0,0)$.

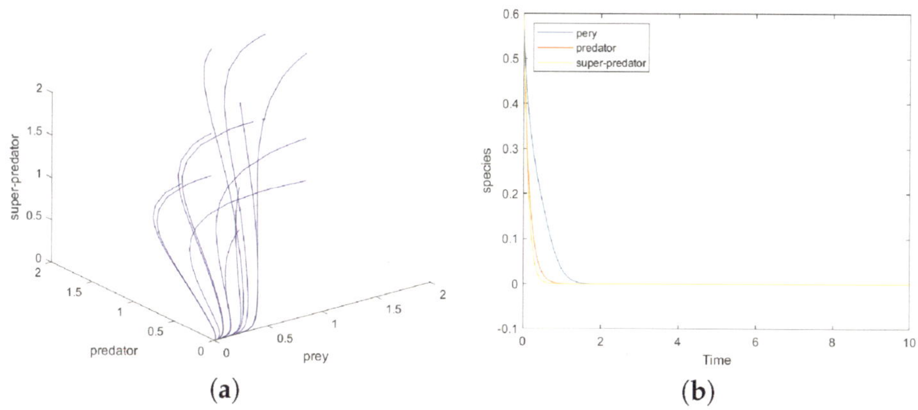

Figure 1.
Phase plot and time series diagram of the model system (3) with set of parameters M_1 = 6.75, M_2 = 5.06, P_1 = 8, P_2 = 10, n = 10, k = 50, a = 59 and other set of parameters in Eq. (12).

Notice: The initial conditions of the LHS graph are defined as [1,1,2]; [1.5,2,2]; [2,1.5,2]; [2,2,2]; [1,1,0.5]; [1.5,2,0.5]; [2,1.5,0.5]; [2,2,0.5]; [1,1,1]; [1.5,2,1]; [2,1.5,1] and [2,2,1]. The initial conditions of the RHS graph are defined as [0.6,0.6,0.6].

We can find in **Figure 1** that the number of all three species rapidly tends to 0 after a certain period of time (all three species are trending extinct). We have proved that no bifurcation situation exists around $C_1(0,0,0)$.

6.2.2 Type 2 $C_2(1,0,0)$

Figure 2 show the stability of point $C_2(1,0,0)$ in relation to the three species based on time t under the corresponding conditions. The number of predators and super-predators tends to close to zero, while the number of preys tends to a positive value (equals to 1) along with the change of time t. In particular, the number of predators tends to increase in previous short period of time and then tends to be decreasing, while the number of super-predators tends to a simple trend of gradually decreasing until close to 0, then stays stable. The number of predators tends to decrease for a short period of time, then increase until the value closes to 1 and remain stable (the conversion rate of the super-predator e_1 is greater than 0). By the conclusion above, no transcritical bifurcation exists around $C_2(1,0,0)$.

6.2.3 Type 3 $C_3(\frac{a}{k},0,0)$

Figure 3 demonstrated the analytical conclusion that $C_3(\frac{a}{k},0,0)$ is an equilibrium point and shown the stability around $C_3(\frac{a}{k},0,0)$.

Notice: The initial conditions of the LHS graph are defined as [2,0.5,2]; [1.5,2,2]; [2,1.5,2].

Figure 3 illustrates that the number of predators and super-predators tends to 0 while the number of prey tends to a positive constant $\frac{a}{k}$. There is no significant fluctuation compared with **Figure 2**. All three species maintained a decreasing trend until their populations stabilized. Specifically, the product of consumption rate and consumption rate of prey, m_1e_1, is less than the death rate of predator d_1 (since by solving the Jacobian

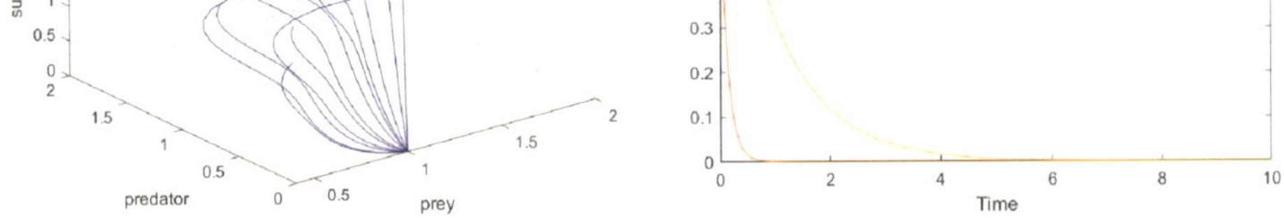

(a) (b)

Figure 2.
*Phase plot and time series diagram of the model system (3) with set of parameters M_1 = 5, M_2 = 4.5, P_1 = 7, P_2 = 1, n = 20, k = 200, a = 1 and other set of parameters in Eq. (12). Initial conditions are defined same as **Figure 1** in this model system (3).*

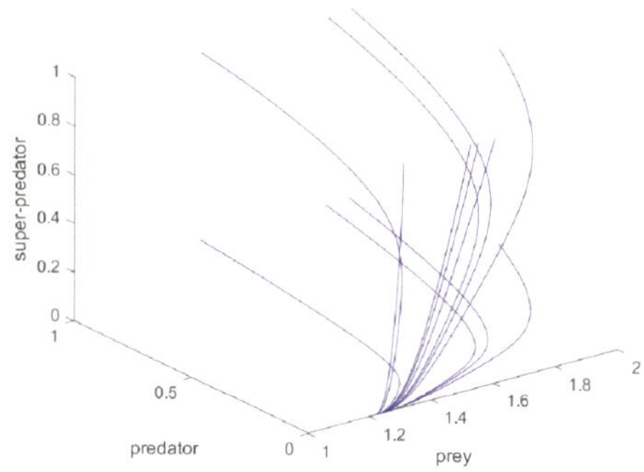

(a)

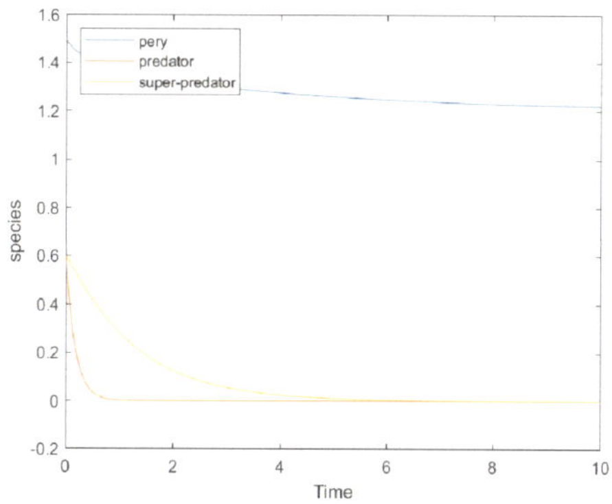

(b)

Figure 3.
Phase plot and time series of the model system (3) with set of parameters M_1 = 0.3, M_2 = 0.3, P_1 = 6, P_2 = 0.8, n = 10, k = 50, a = 60 and other set of parameters in Eq. (12). The initial conditions of the RHS graph are defined as [1.5, 0.6, 0.6] in the model system (3).

matrix, we have two boundary conditions: $M_1 a < P_1 k$ and $a > k$. Hence, we can get $m_1 e_1 < d_1$). It illustrates that the population of predators is dying faster than growing, while the number of individuals of prey that are not consumed (survive) increases.

Likewise, by analyzing the bifurcation conditions, the parameter range does not fit the constrained condition (which we have analyzed that $m_1 e_1 < d_1$). Hence, there is no transcritical bifurcation and Hopf-bifurcation exists around $C_3\left(\frac{a}{k},0,0\right)$.

6.2.4 Type 4 $C_4\left(\frac{P_1}{M_1},\frac{f\left(\frac{P_1}{M_1}\right)}{M_1},0\right)$

By solving the equation of nondimentionalized system (The partial differential equations of x based on t equal to 0), we can get one possible point $C_4\left(\frac{P_1}{M_1},\frac{f\left(\frac{P_1}{M_1}\right)}{M_1},0\right)$.

Specifically, $\frac{f\left(\frac{P_1}{M_1}\right)}{M_1}$ is a function where $f\left(\frac{P_1}{M_1}\right)=\frac{\left(1-\frac{P_1}{M_1}\right)\left(\frac{P_1}{M_1}k-a\right)}{\frac{P_1}{M_1}k+n}$ (where parameters $P_1,\ M_1,\ a,\ k,\ n$ are known). And we will do some numerical analysis based on our stable point plot, **Figure 4**, which fits this situation.

Notice: The initial conditions of the LHS graph are defined as (1,1,2)(1.5,2,2)(2,1.5,2)(0.5,1,2)(1,1,2.2)(1.5,2,2.2)(2,1.5,2.2)(2,2,2.2)(1,1,1)(1.5,1,2)(2,1.5,2.2)(2,1.5,2). The initial conditions of the RHS graph are defined as (0.6,0.6,0.6).

Figure 4 demonstrated that the population of prey and predator shows obvious fluctuations in the early stage (time t within [0,100]). According to the figure, the fluctuations are approximately periodic (with period approximately t = 30). The population fluctuates in a Violent amplitude at first, then gradually slows down, until it reaches the equilibrium value ($x=\frac{P_1}{M_1}, y=\frac{f\left(\frac{P_1}{M_1}\right)}{M_1}=\frac{\left(1-\frac{P_1}{M_1}\right)\left(\frac{P_1}{M_1}k-a\right)}{M_1\left(\frac{P_1}{M_1}k+n\right)}$). The number of super-predator is decreasing with a high tangent value. This is because the

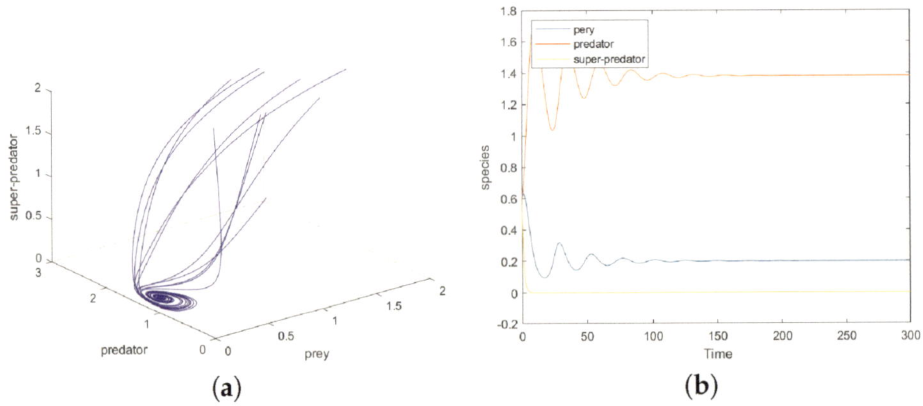

Figure 4.
Phase plot and time series diagram of the model system (3) with set of parameters $M_1 = 0.5$, $M_2 = 0.2$, $P_1 = 0.1$, $P_2 = 1$, $n = 1$, $k = 100$, $a = 1.9$ and other set parameter values same as Eq. (12).

super-predator death rate d_2 is much greater than the product of the conversion rate of super-predator e_2 and the consumption rate from predator to super-predator m_2 $(m_2e_2 = 0.04 < d_2 = 0.2)$.

Here we calculated the parameter values of Hopf-bifurcation point using the Newton iterative method based on the equation: $1 + \frac{a}{k} - 2\frac{P_1}{M_1^{[hb_1]}} - \frac{\left(P_1 - M_1^{[hb_1]}\right)\left(P_1k - aM_1^{[hb_1]}\right)}{\left(P_1k + nM_1^{[hb_1]}\right)M_1^{[hb_1]}} = 0.$

6.2.5 *Type 5 and type 6 $C_{5,6}(u, v, w)$*

Next, we will carry on studying the stability and several plots that show the relation between three species based on time during two different conditions of the point $C_5(u, v, w)$. Based on our analytical studies, we have obtained a series of equations about u, v and w. Solving the equation, we get:

$$u = u^* = \frac{(aM_2 + kM_2 - M_1P_2k) \pm \sqrt{(aM_2 + kM_2 - M_1P_2k)^2 + 4kM_2(M_1P_2n - M_2a)}}{2kM_2}$$

$v = \frac{P_2}{M_2}$ and $w = \frac{M_1u^* - P_1}{M_2}$. Specially, $C_6(u, v, w)$ and $C_5(u, v, w)$ are essentially points of the same type of (u, v, w). For the convenience of illustrating and comparing the different properties of the two points we name the second point as type 6: $C_6(u, v, w)$.

6.2.5.1 *Type 5 $C_5(u, v, w)$*

The first conditions are exhibited in **Figure 5**:

Since **Figure 5** looks like $C_4\left(\frac{P_1}{M_1}, \frac{f\left(\frac{P_1}{M_1}\right)}{M_1}, 0\right)$-type point ($w$ is a small number and close to 0), we put extra two plots. Here we can obviously see the equilibrium point in a $C_{5,6}(u, v, w)$ type point.

From **Figures 5–7**, we can clearly observe that all three species will eventually remain stable in different positive values as time t extends. The value of the number of prey first

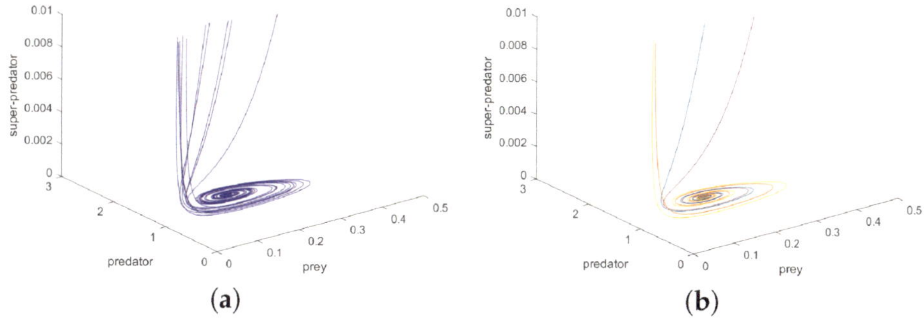

Figure 5.
*Magnified view of **Figure 4** showing the model system (3) with set of parameters in Eq. (12).*

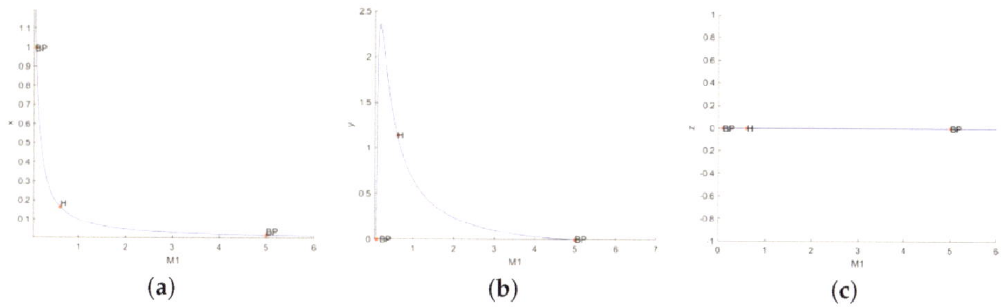

Figure 6.
Shows the bifurcation situation for prey, predator and super-predator. Fixed other parameters, there exists one Hopf-bifurcation in the model system (3).

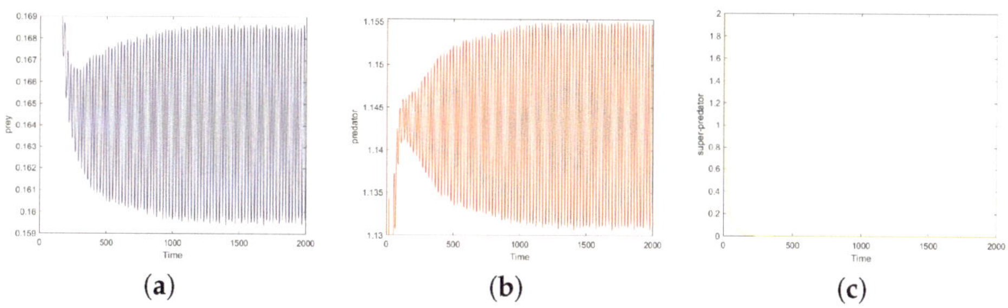

Figure 7.
Hopf-bifurcation around C_4 for parameter values $M_1 = 0.601, P_1 = 0.25$: $m_1 = 0.061, m_2 = 0.2, e_1 = 0.08, e_2 = 0.1, r = 0.8, d_1 = 0.08, d_2 = 0.2$. Here, we use Newtonian iteration and Matcont to verify M_1 values in the model system (3).

increases and then changes to decreases, accompanied by a small fluctuation. The value of predators increases and then decreases, also accompanied by fluctuates slight. The number of super-predators has been decreasing with a slight fluctuation until it reached a positive value $\frac{M_1 u * - P_1}{M_2}$ and has remained horizontally parallel.

In **Figure 8**, the phase portrait of the Hopf-bifurcation around C_4, and started from initial point (0.2,0.2,0.4) and (1,1,2) in the model system (3). We can obviously notice that there should exist a Hopf bifurcation around $C_5(u, v, w)$ theoretically.

While the Hopf-bifurcation is at the corresponding situation where $z < 0$. Hence, we say Hopf-bifurcation does not exist around $C_5(u, v, w)$. However, it does not mean Hopf-bifurcation does not exist around all (u, v, w), according to our proof. In this case, we decided to use different parameters to try.

6.2.5.2 *Type 6* $C_6(u, v, w)$

Finally, we are going to study the stability around the point $C_6(u, v, w)$ and the result of plots related to three species over time under another condition (**Figure 9**). Specifically, $C_6(u, v, w)$ choose other parameter values that are different from $C_6(u, v, w)$.

Comparing **Figure 10** with **Figure 9**, the result can illustrate that under the conditions of **Figure 11**, the number of species fluctuated more acutely, and tends to be slower to reach the equilibrium value. In addition, the equilibrium values of prey and predators are greater than the value in the initial condition after fluctuating, while the equilibrium value of the super-predator is smaller compared to the value in the initial condition (the number of prey and predators increases while the number of super-predators decreases).

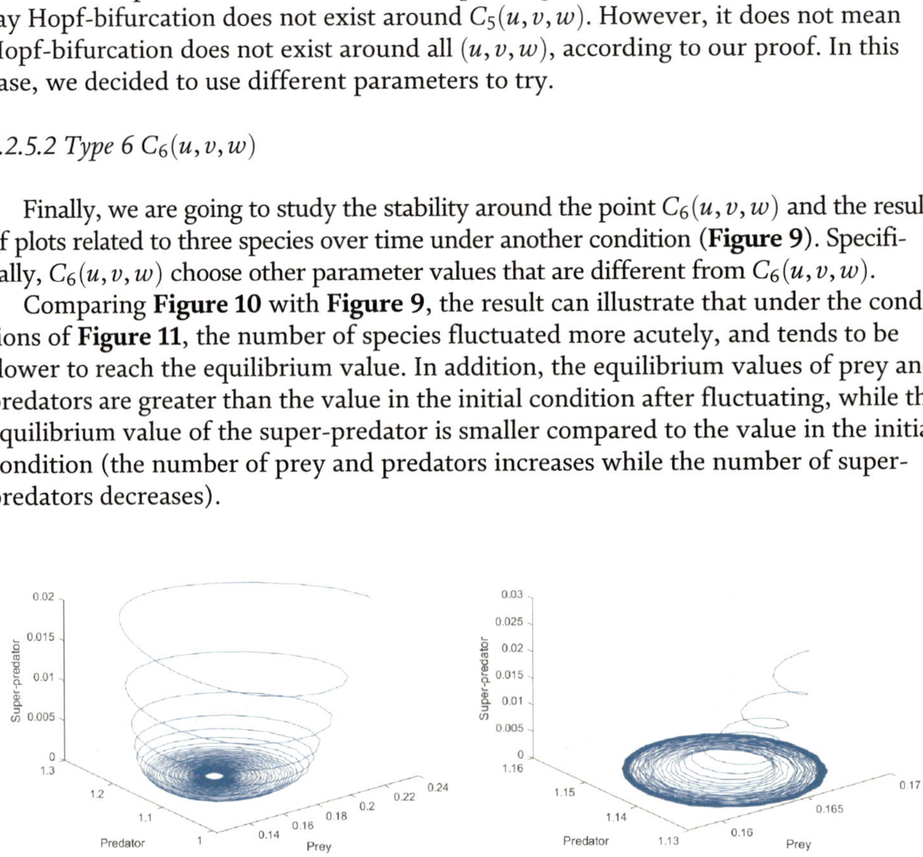

(a) (b)

Figure 8.
Phase portrait of Hopf-bifurcation around started from initial point $(0.2, 0.2, 0.4)$ *and* $(1, 1, 2)$ *in the model system* (3).

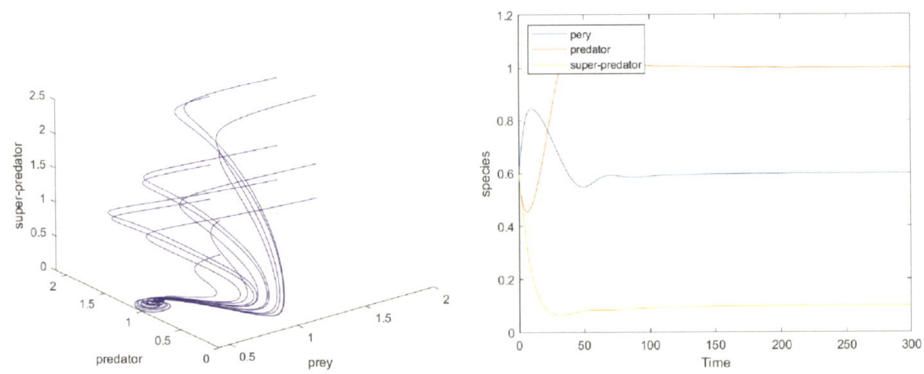

(a) (b)

Figure 9.
In condition of parameter: $M_1 = 0.2; M_2 = 0.2; P_1 = 0.1; P_2 = 0.2; n = 10; k = 50; a = 10$ $(m1 = 0.02, m2 = 0.2, e1 = 0.16, e2 = 0.8, r = 0.8, d1 = 0.08, d2 = 0.16)$. *The initial conditions are defined the same as* ***Figure 1*** *in the model system 3.*

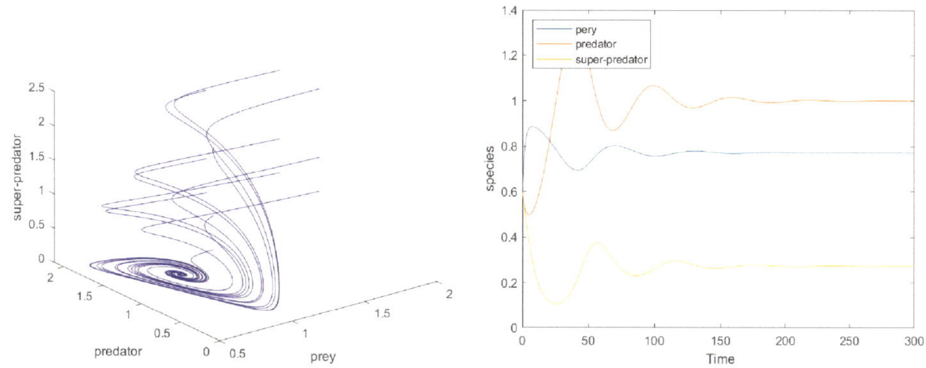

(a) (b)

Figure 10.
In condition of parameter: $M_1 = 0.2; M_2 = 0.2; P_1 = 0.1; P_2 = 0.2; n = 10; k = 200; a = 10$ $(m1 = 0.016, m2 = 0.2, e1 = 0.05, e2 = 0.08, r = 0.8, d1 = 0.08, d2 = 0.16)$. *The initial conditions are defined the same as* ***Figure 1*** *in the model system 3.*

Fixed the value of two predator $d_1 = 0.08$ and $d_2 = 0.16$. For the type 5, we have $m_1e_1 = 0.02 \times 0.16 = 0.0032$, $m_2e_2 = 2 = 0.02 \times 0.1 = 0.002$, which means $m_1e_1 > m_2e_2$. For the type 6, we have $m_1e_1 = 0.016 \times 0.05 = 0.0008$, $m_2e_2 = 2 = 0.02 \times 0.08 = 0.016$, which means $m_1e_1 < m_2e_2$. According to the meaning of these parameters, we can see the reason why the equilibrium point in **Figure 11** (w is approximately equal to 0.25) is higher than the equilibrium point in **Figure 12** (w is around 0.09). The rate between m_2e_2 and m_1e_1 means the super-predator consumes the predator. The larger the rate, the more the super-predator population remains. Furthermore, compared to **Figure 9**, the super-predator experienced an obvious fluctuation in **Figure 10**. In order to demonstrate the results of our model research, we calculated and re-selected the appropriate parameter values based on our result equation above to fit the Hopf-bifurcation condition, and used Newton's iterative method to estimate the values of each parameter.

We can observe that x, y, z fluctuates periodically (forming a ring) within a stable interval, respectively. Furthermore, we have shown the existence of a limit point (LP) in **Figures 13** and **14**. Now we re-selected the appropriate parameter and plotted our numerical findings below.

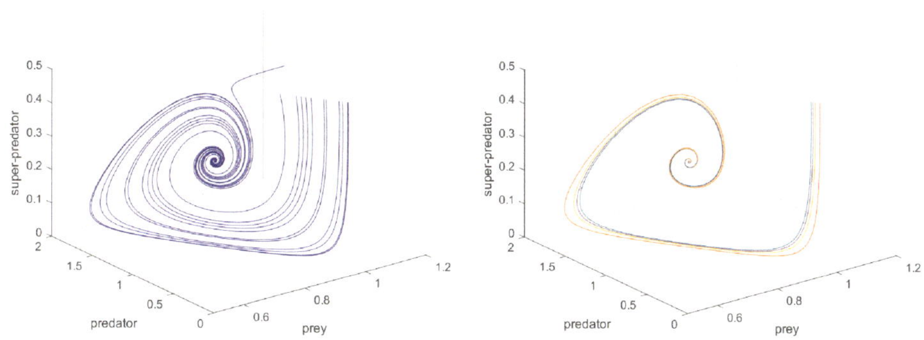

(a) (b)

Figure 11.
Magnified view of ***Figure 10*** *in the model system 3.*

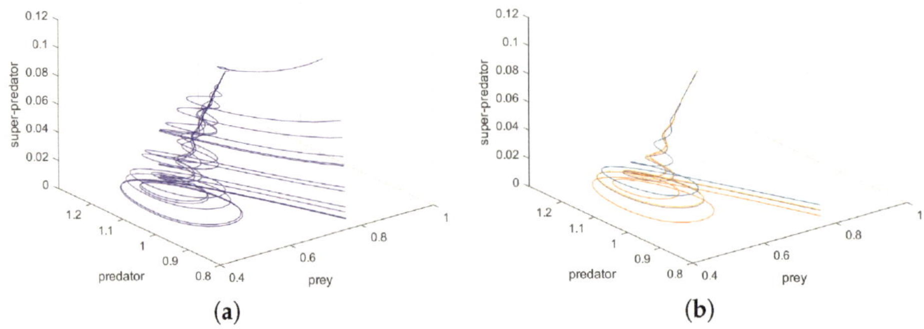

Figure 12.
his figure is a magnified view of ***Figure 9*** *in the model system 3.*

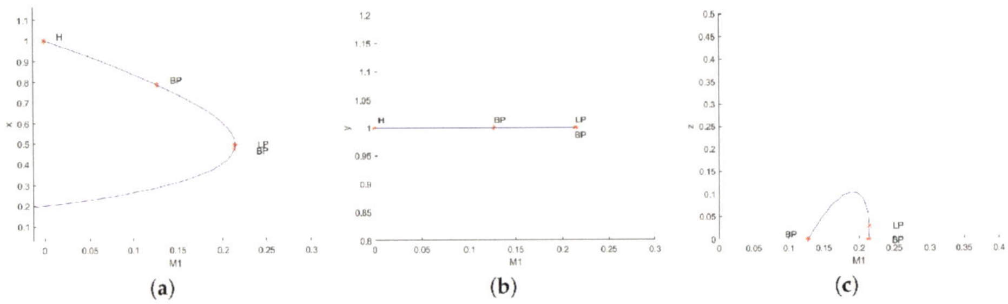

Figure 13.
Shows the bifurcation situation for prey, predator and super-predator in the model system 3.

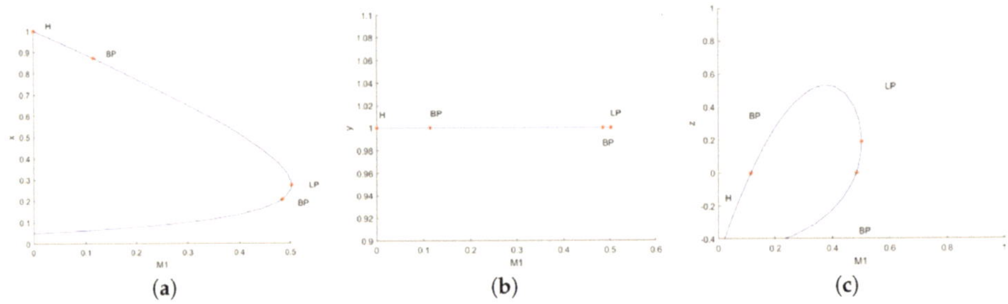

Figure 14.
Shows the bifurcation situation for prey, predator and super-predator. Fixed other parameters, there exists no Hopf-bifurcation point (H) ($z < 0$) and one limit point (LP) with different values of M_1, the same conclusion as $C_5(u, v, w)$ in the model system 3.

7. Discussion and conclusion

In this chapter, we first mentioned in the introduction that food chain research has long been an important biological topic, and mathematical modeling is an important tool for tackling biological ecological problems. At the same time, we discussed the origins of Allee effects as well as the biological importance of both the strong and mild Allee effects. We discovered that the Allee effect is currently a hot topic, so we decided to look further into it.

Then, the system of the Eq. (3) has been nondimensionalized according to $u = \frac{x}{k}, v = \frac{y}{ke_1}, \ w = \frac{z}{ke_1e_2} \tau = rt$. We discussed about different origins $C_1(0,0,0), C_2(1,0,0), C_3(\frac{a}{k},0,0), C_4(u_1, v_1, 0), C_5(u, v, w)$ and the different behaviors of this system around them including system local stability, global stability, and bifurcation analysis. About local stability, we found the condition under when this hyperbolic equilibrium point is locally asymptotically stable. If the eigenvalues of one system are all real number, it does not have saddle-node, Hopf-bifurcation. In C_1 , C_2 *and* C_3, it does not have saddle-node, Hopf-bifurcation, they cannot experience transcritical bifurcation. Meanwhile, we evaluated the global stability condition $(C_2 : M_1 < P_1, \ C_3 : \ a > k, \ aM_1 < kP_1)$. In C_4, the system enters into a Hopf-bifurcation around C_4 when $\frac{u}{ku+n}\left[\frac{u}{(ku+n)^3}(n+a)(k+n) - (uk+n)^2\right]^2 - 4M(1-u)(uk-a) < 0$ In the hyperbolic equilibrium point C_4, we use the Routh-Hurwitz theorem to evaluate the condition of local asymptotic stability in the region of $\left[\frac{n}{(ku_1+n)^2}(1-u_1)(ku_1-a) + \frac{u_1}{ku_1+n}(k+a-2u_1k)\right] - M_1v_1 < 0, \ v_1 < \frac{P_2}{M_2}$. The system (3) becomes locally asymptotically stable around C_5 if it satisfies the conditions $\frac{P_1}{M_1} < u < \frac{P_1}{M_1 + m_{11}\frac{M_2^2}{M_2}}$ and $m_{11} < 0$ where $m_{11} = \frac{\left[\frac{P_2M_1k}{M_2} - (k+a)\right]u + 2\left(a + \frac{nP_2M_1}{M_2}\right)}{uk+n}$. Also, this system experiences a Hopf-bifurcation around $M_1 = M_1^{[hb_2]}$ where $M_1^{[hb_2]}$ satisfy the equation $\left(\frac{m_{11}\left(M_1^{[hb_2]}\right)^2}{M_2} + M_1^{[hb_2]}\right)u - P_1 = 0$. Then, we evaluate five equilibrium points of this system in different situations. First of all, in C_1, a is significantly greater than k in the value which goes against true situation. In the next points $C_2(1,0,0)$ and $C_3(\frac{a}{k},0,0)$, both these two equilibrium points indicate that the prey approaches a stable value while the predators become extinct. In the C_3 prime s graphs, we find that the prey and predator approach to stable quantity, while the super-predator approaches zero which implies that extinction of predator population is possible for other model but not our model. Also, we think this case is quiet similar to actual case for nature. Furthermore, we plotted the bifurcation graph via matcont to conclude that the population of prey and predator fluctuates within each certain stable integral after a period of time. In equilibrium point (x,y,z), we got some plot which implies that all the three species will remain a certain number of population. Based on our whole process, we find that the value of a and k play an important factor on the function of Alee effect for this system. This study underscores the significance of incorporating Allee effects into ecological models, given their substantial impact on species persistence and extinction dynamics. By capturing the influence of low population densities, the Allee effect introduces critical thresholds that alter system stability and long-term behavior. Furthermore, the work emphasizes the importance of nonlinear interactions and suggests that future investigations should focus on exploring potential chaotic dynamics in similar systems to better understand the complexity of ecological processes. Future work could incorporate stochasticity and spatial dynamics to more accurately simulate realistic ecological processes. Incorporating spatial heterogeneity allows for the examination of pattern formation, species dispersal, and localized interactions that are often critical in natural systems. Building on this, our ongoing research focuses on the role of spatial dynamics and the development of amplitude equations to capture the emergence and evolution of spatiotemporal patterns in ecological models influenced by the Allee effect.

Author contributions

The author has concluded manuscript writing, model formulation and simulation, and final manuscript.

Conflict of interest

The authors declare that there is no conflict of interest regarding the publication of this chapter.

Author details

Sarita Kumari*† and J. Omana†
Vellore Institute of Technology, Chennai, India

*Address all correspondence to: sarita.kumari@vit.ac.in

† These authors contributed equally.

References

[1] Haque M, Ali N, Chakravarty S. Study of a tri-trophic prey-dependent food chain model of interacting populations. Mathematical Biosciences. 2013;**246**(1): 55-71

[2] Pattanayak D, Mishra A, Dana SK, Bairagi N. Bistability in a tri-trophic food chain model: Basin stability perspective. Chaos: An Interdisciplinary Journal of Nonlinear Science. 2021;**31**(7)

[3] Murray JD. Mathematical Biology: I. An Introduction. Vol. 17. Springer Science & Business Media; 2007

[4] Berryman AA. The orgins and evolution of predator-prey theory. Ecology. 1992;**73**(5):1530-1535

[5] Pal PJ, Mandal PK, Lahiri KK. A delayed ratio-dependent predator–prey model of interacting populations with Holling type III functional response. Nonlinear Dynamics. 2014;**76**:201-220

[6] Holling CS. Some characteristics of simple types of predation and parasitism. The Canadian Entomologist. 1959;**91**(7): 385-398

[7] Holling CS. The components of predation as revealed by a study of small-mammal predation of the European Pine Sawfly. The Canadian Entomologist. 1959;**91**(5):293-320

[8] Misra AK, Dubey B. A ratio-dependent predator-prey model with delay and harvesting. Journal of Biological Systems. 2010;**18**(02):437-453

[9] Arditi R, Ginzburg LR. Coupling in predator-prey dynamics: Ratio-dependence. Journal of Theoretical Biology. 1989;**139**(3):311-326

[10] Sivasamy R, Sivakumar M, Balachandran K, Sathiyanathan K. Spatial pattern of ratio-dependent predator-prey model with prey harvesting and cross-diffusion. International Journal of Bifurcation and Chaos. 2019;**29**(03):1950036

[11] Gakkhar S, Naji RK. Order and chaos in predator to prey ratio-dependent food chain. Chaos, Solitons & Fractals. 2003; **18**(2):229-239

[12] Hsu SB, Hwang TW, Kuang Y. A ratio-dependent food chain model and its applications to biological control. Mathematical Biosciences. 2003;**181**(1): 55-83

[13] Pal PJ, Mandal PK. Bifurcation analysis of a modified Leslie–Gower predator–prey model with Beddington–DeAngelis functional response and strong Allee effect. Mathematics and Computers in Simulation. 2014;**97**: 123-146

[14] Beddington JR. Mutual interference between parasites or predators and its effect on searching efficiency. The Journal of Animal Ecology. 1975: 331-340

[15] DeAngelis DL, Goldstein RA, O'Neill RV. A model for tropic interaction. Ecology. 1975;**56**(4):881-892

[16] Crowley PH, Martin EK. Functional responses and interference within and between year classes of a dragonfly population. Journal of the North American Benthological Society. 1989; **8**(3):211-221

[17] Sivasamy R, Sathiyanathan K, Balachandran K. Dynamics of a modified Leslie-Gower model with Crowley-Martin functional response and prey harvesting. Journal of Applied Nonlinear Dynamics. 2019;**8**(4):621-636

[18] Li Y, Rafaqat M, Zia TJ, Ahmed I, Jung CY. Flip and Neimark-Sacker bifurcations of a discrete time predator-pre model. IEEE Access. 2019;**7**: 123430-123435

[19] Lin Y, Din Q, Rafaqat M, Elsadany AA, Zeng Y. Dynamics and chaos control for a discrete-time Lotka-Volterra model. IEEE Access. 2020;**8**: 126760-126775

[20] Hastings A, Powell T. Chaos in a three-species food chain. Ecology. 1991; **72**(3):896-903

[21] Jana D, Agrawal R, Upadhyay RK. Top-predator interference and gestation delay as determinants of the dynamics of a realistic model food chain. Chaos, Solitons & Fractals. 2014;**69**:50-63

[22] Pathak S, Maiti A, Samanta GP. Rich dynamics of a food chain model with Hassell-Varley type functional responses. Applied Mathematics and Computation. 2009;**208**(2):303-317

[23] Tanner JT. The stability and the intrinsic growth rates of prey and predator populations. Ecology. 1975; **56**(4):855-867

[24] Upadhyay RK, Naji RK. Dynamics of a three species food chain model with Crowley–Martin type functional response. Chaos, Solitons & Fractals. 2009;**42**(3):1337-1346

[25] Shah K, Alqudah MA, Jarad F, Abdeljawad T. Semi-analytical study of Pine Wilt Disease model with convex rate under Caputo-Febrizio fractional order derivative. Chaos, Solitons & Fractals. 2020;**135**:109754

[26] Shah K, Jarad F, Abdeljawad T. On a nonlinear fractional order model of Dengue fever disease under Caputo-Fabrizio derivative. Alexandria Engineering Journal. 2020;**59**(4): 2305-2313

[27] Samina, Shah K, Khan RA. Stability theory to a coupled system of nonlinear fractional hybrid differential equations. Indian Journal of Pure and Applied Mathematics. 2020;**51**:669-687

[28] Vivek D, Shah K, Kanagarajan K. Dynamical analysis of Hilfer-Hadamard type fractional pantograph equations via successive approximation. Journal of Taibah University for Science. 2019; **13**(1):225-230

[29] Lopez-Ruiz R, Fournier-Prunaret D. Indirect Allee effect, bistability and chaotic oscillations in a predator-prey discrete model of logistic type. Chaos, Solitons & Fractals. 2005;**24**(1):85-101

[30] Putnik GD, Cruz-Cunha MM, editors. Encyclopedia of Networked and Virtual Organizations. IGI Global; 2008

[31] Gilani SA, Din Q. Dynamics of a Leslie–Gower predator–prey model with Allee effect in predator: Stability, bifurcations, and pattern formation. International Journal of Dynamics and Control. 2025;**13**(7):269

Chapter 4

Controlling the Dynamics of Soliton Molecules in Binary Bose-Einstein Condensates

Maïk Delon Mboumba, Thierry Blanchard Ekogo and Alain-Brice Moubissi

Abstract

The analogy between nonlinear optics and Bose-Einstein condensates (BECs) is a fascinating area of study due to the similarities in their underlying physical principles. Both systems are governed by similar nonlinear partial differential equations, which allow researchers to draw parallels between the behaviors observed in these two distinct fields. Among the phenomena observed, the soliton is of particular interest. In this chapter, we demonstrate the possibility of controlling the formation and dynamics of soliton molecules in binary dipolar Bose-Einstein condensates (BECs) with spin-orbit (SO) coupling. The study is carried out analytically using the variational approach and validated numerically through split-step Fourier simulations. Our results reveal that the integrity and stability of the soliton molecules are highly sensitive to the system's physical parameters, particularly the interplay between local (contact) and nonlocal (dipolar) interactions in the presence of SO coupling. By appropriately tuning these interactions, one can effectively control the binding, oscillatory dynamics, and potential dissociation of the soliton molecules.

Keywords: coupled dipolar Bose-Einstein condensates, soliton molecules, spin-orbit coupling, variational approach, intra-component interactions, inter-component interactions

1. Introduction

The exploration of ultracold atomic physics beyond conventional Bose-Einstein condensates (BECs) has emerged as one of the most active and compelling frontiers in modern atom optics [1]. The idea of a Bose-Einstein condensate was first proposed by Albert Einstein [2], building on the work of Satyendra Nath Bose [3], who had developed statistical rules for photons. It was not until 1995 that the first BEC was experimentally created by Eric Cornell, Carl Wieman [4], and later Wolfgang Ketterle in rubidium atoms [5]. This achievement earned them the Nobel Prize in Physics in 2001. A BEC is a unique state of matter that occurs when a group of bosonic particles (particles with integer spin) are cooled to temperatures extremely close to absolute zero (0 Kelvin or -273°C). At these ultra-low temperatures, the particles lose almost all of their individual energy and begin to occupy the same quantum state, behaving as a single coherent quantum entity. This means that the particles exhibit collective quantum properties that are not observed in other states of matter like solids, liquids, or gases.

Shortly after the first experimental realization of Bose-Einstein condensates (BECs), numerous research groups began to investigate their properties, both theoretically and experimentally [6]. These early studies laid the foundation for understanding a wide range of quantum phenomena in dilute ultracold gases [6]. These investigations include the observation and control of solitons in both single-component and multi-component systems [6–8]. Atomic-gas BECs have proven to be excellent platforms for studying and manipulating nonlinear matter-wave dynamics, owing to their high degree of tunability and coherence [6–8]. Moreover, the manipulation of matter waves can be effectively achieved using a broad range of techniques derived from atomic, molecular, and optical (AMO) physics, such as Feshbach resonances, optical lattices, and spin-orbit coupling schemes [1, 6]. For example, among nonlinear excitation described theoretically and experimentally found, one can include dark solitons [9–11], bright solitons [12–14], different vector solitons, and so on. From a theoretical standpoint, the dynamics of a BEC is well described by the Gross-Pitaevskii eq. (GP) [15, 16]. The (GP) equation is a variant of the nonlinear Schrödinger equation (NLSE) that incorporates an external trapping potential, representing the confinement of the condensate [1]. The effects of interatomic interactions within the condensate are captured by nonlinear terms in the equation. It is well established that two-body and three-body interactions, modeled by cubic and quintic nonlinearities, respectively, play a crucial role in governing the dynamical behavior of the condensate [7, 8]. At the local level, these nonlinearities arise from elastic atom-atom collisions. However, a more detailed description of the system must account for the fact that atoms carry magnetic dipole moments, originating from both electronic and nuclear spins [17, 18]. These dipolar interactions introduce nonlocality into the system, giving rise to long-range dipole-dipole interactions (DDIs), which significantly affect the structure and stability of the condensate [17, 18]. Nonlocal nonlinearity arises in non-alkali atom condensates such as chromium ^{52}Cr [19], dysprosium ^{164}Dy [20], and erbium ^{168}Er [21] due to their large permanent magnetic dipole moment. Dipolar BECs present many unusual properties not found in BECs with just local interactions [22–24]. Another topic of significant current interest is spin-orbit (SO) coupling [25]. In quantum physics, SO coupling refers to the interaction between a particle's spin degree of freedom and its orbital motion [26, 27]. This coupling plays a central role in various quantum systems, from atomic and condensed matter physics to ultracold atomic gases [26, 27]. In ultracold quantum gases, the SO coupling can be synthetically engineered using two counter-propagating Raman lasers that couple two hyperfine ground states [25–27]. In the context of Bose-Einstein condensates, SO coupling introduces spin-dependent dispersion, leading to a rich variety of novel quantum phases, anisotropic excitations, and complex nonlinear dynamics, including the formation of spinor solitons and topological structures [25–27]. In combination with the local nonlinearity due to the atom-atom interactions, the SO coupling provides a platform for the study of various patterns and collective excitations in BECs [25–27].

This chapter investigates the formation and dynamics of soliton molecules in binary dipolar Bose-Einstein condensates (BECs) within the framework of the coupled nonlocal Gross-Pitaevskii equations. Specifically, we analyze the time evolution of the amplitudes and widths of the individual components of the soliton molecules using the variational approximation. The analytical predictions derived from this method are systematically compared with the results obtained through numerical simulations based on the split-step Fourier method, demonstrating good quantitative agreement between the two approaches.

We aim to meet the needs of both newcomers and experts in the field. To support readers who are new to the subject, Section 2 of this chapter provides a tutorial-style introduction to the nonlinear dynamics of Bose-Einstein condensates (BECs). This section includes a brief overview of the derivation of the Gross-Pitaevskii equation and outlines the main research motivations for studying dipolar interactions and spin-orbit coupling in BECs. In Section 3, we present a systematic and comprehensive theoretical framework for the system under investigation. We introduce the governing model, based on coupled nonlocal Gross-Pitaevskii equations, and describe the variational approximation technique used to analyze the dynamics. Section 4 is devoted to the analytical and numerical investigation of soliton molecule dynamics, highlighting the interplay between local and nonlocal interactions as well as the role of spin-orbit coupling. Finally, Section 5 summarizes the main findings and presents the conclusions of the chapter.

2. Nonlinear dynamics of Bose-Einstein condensates

We provide in this section a brief overview of the general mathematical description of a weakly interacting quantum gas of BEC and present the main research motivations concerning soliton in dipolar BECs and spin-orbit coupling in BECs.

2.1 The Gross-Pitaevskii equation

The Gross-Pitaevskii equation, so named after Eugene P. Gross and Lev Petrovich Pitaevskii, is the equation that describes the state and dynamics of an ultracold gas of identical bosons, i.e., the Bose-Einstein condensate. This equation is based on the mean-field approximation and the model of pseudo-potential interaction. The theory consists of the following assumptions [1, 7]:

- A large fraction of atoms are in the same state.
- The potential describing the interaction between atoms is replaced by a pseudo-potential which accounts for a contact binary collision.

The many-body hamiltonian describing a system of N interacting identical bosons of mass m, and immersed in an external trapping potential V_{ext} is given, in the second quantization, by

$$\begin{aligned}\hat{H} = &\int d\mathbf{r}\hat{\Psi}^{\dagger}(\mathbf{r},t)\left[-\frac{\hbar^2}{2m}\nabla^2 + V_{ext}(\mathbf{r},t)\right]\hat{\Psi}(\mathbf{r},t) \\ &+\frac{1}{2}\int\int d\mathbf{r}d\mathbf{r}'\hat{\Psi}^{\dagger}(\mathbf{r},t)\hat{\Psi}^{\dagger}(\mathbf{r}',t)V(\mathbf{r}-\mathbf{r}')\hat{\Psi}(\mathbf{r}',t)\hat{\Psi}(\mathbf{r},t),\end{aligned} \tag{1}$$

where $-\frac{\hbar^2}{2m}\nabla^2$ is the kinetic energy operator. $V(\mathbf{r}-\mathbf{r}')$ is the exact potential for two-body interaction. $\hat{\Psi}^{\dagger}(\mathbf{r},t)$ and $\hat{\Psi}(\mathbf{r},t)$ are the creation and annihilation operators representing the addition or subtraction of boson at point r in space in time t, respectively. These operators satisfy commutation relations

$$\left[\hat{\Psi}^{\dagger}(\mathbf{r},t),\hat{\Psi}^{\dagger}(\mathbf{r}',t)\right]=\left[\hat{\Psi}(\mathbf{r},t),\hat{\Psi}(\mathbf{r}',t)\right]=0,\quad \left[\hat{\Psi}(\mathbf{r},t),\hat{\Psi}^{\dagger}(\mathbf{r}',t)\right]=\delta(\mathbf{r}-\mathbf{r}'). \quad (2)$$

Using Heisenberg's relation $i\hbar\partial\hat{\Psi}/\partial t=\left[\hat{\Psi},\hat{H}\right]$ and the commutation relations, we obtain the equation of motion

$$i\hbar\frac{\hat{\Psi}(\mathbf{r},t)}{\partial t}=\left[-\frac{\hbar^2}{2m}\nabla^2+V_{ext}(\mathbf{r},t)\right]\hat{\Psi}(\mathbf{r},t)+\int d\mathbf{r}'\hat{\Psi}^{\dagger}(\mathbf{r}',t)V(\mathbf{r}-\mathbf{r}')\hat{\Psi}(\mathbf{r}',t)\hat{\Psi}(\mathbf{r},t). \quad (3)$$

Since Bose-Einstein condensation involves a macroscopic population in a single quantum state, it is appropriate to consider a mean-field approach. In the Bogoliubov approximation, the idea is to separate the bosonic field into two parts such that

$$\hat{\Psi}(\mathbf{r},t)=\psi(\mathbf{r},t)+\delta\psi(\mathbf{r},t). \quad (4)$$

The first $\psi(\mathbf{r},t)$ is a complex function corresponding to the macroscopic occupation of the ground state (condensed part); the second part $\delta\psi(\mathbf{r},t)$ describes the non-condensed fraction corresponding to quantum and thermal fluctuations. In other words, $\psi(\mathbf{r},t)=<\hat{\Psi}(\mathbf{r},t)>$ is the mean value of the bosonic field operator and $\delta\psi(\mathbf{r},t)$, the deviation around this mean. At very low temperatures, quantum fluctuations in the system can be neglected. Thus, replacing Eq. (4) in Eq. (3) and neglecting $\delta\psi(\mathbf{r},t)$, we obtain the Schrödinger equation for the condensate wave function

$$i\hbar\frac{\psi(\mathbf{r},t)}{\partial t}=\left[-\frac{\hbar^2}{2m}\nabla^2+V_{ext}(\mathbf{r},t)\right]\psi(\mathbf{r},t)+\int d\mathbf{r}'\psi(\mathbf{r}',t)V(\mathbf{r}-\mathbf{r}')\psi(\mathbf{r}',t)\psi(\mathbf{r},t). \quad (5)$$

The next step after the Bogoliubov approximation is the simplification of the interatomic contact potential $V(\mathbf{r}-\mathbf{r}')$. On the one hand, this potential has a long-range attractive part, corresponding to the Van der Waals forces between two induced dipoles. On the other hand, a short-range repulsive part, corresponding to the Lennard-Jones type potential, prevents the electronic clouds of two atoms from interpenetrating. To describe the properties of atomic gas, it is not necessary to take into account the precise form of this potential. At very low temperatures, the medium is diluted, and the inter-particle distances are large compared to the range of the potential $V(\mathbf{r}-\mathbf{r}')$. In other words, the order of magnitude of the quantum delocalization of an atom, at temperature T, is given by its thermal wavelength $\lambda_{th}=h/\sqrt{2\pi mk_BT}$. When this becomes greater than the range of the interaction potential, the atom can no longer see the details. We can thus replace the potential $V(\mathbf{r}-\mathbf{r}')$ in (5) by

$$V(\mathbf{r}-\mathbf{r}')=g\delta(\mathbf{r}-\mathbf{r}'). \quad (6)$$

The interaction strength between two particles at low energies is characterized by the parameter $g=4\pi\hbar^2a_s/m$, where a_s is the s-wave scattering length, $\hbar$ is the reduced Planck constant, and m is the atomic mass. The sign of g determines the nature of the interaction: $g>0$ corresponds to repulsive interactions, while $g<0$ indicates attractive interactions. These low-energy collisions are well-described by the s-wave scattering approximation and are effectively parameterized by the single quantity a_s, which is largely independent of the detailed form of the interatomic potential. By replacing Eq. (6) in Eq. (5), we obtain

$$i\hbar\frac{\psi(\mathbf{r},t)}{\partial t}=\left[-\frac{\hbar^2}{2m}\nabla^2+V_{ext}(\mathbf{r},t)\right]\psi(\mathbf{r},t)+g|\psi(\mathbf{r},t)|^2\psi(\mathbf{r},t). \quad (7)$$

Eq. (7) is known as the Gross-Pitaevskii equation (GPE) [1]. It allows us to describe the state and dynamics of a Bose-Einstein condensate at zero temperature. Furthermore, if the external potential is independent of time and the system is in a stationary state, we can separate the spatial dependence and thus obtain the time-independent GPE. This is done by inserting $\psi(\mathbf{r},t)=\psi(\mathbf{r})\,exp\,(-i\mu t/\hbar)$ in Eq. (7), such as

$$\left[-\frac{\hbar^2}{2m}\nabla^2+V_{ext}(\mathbf{r})+g|\psi(\mathbf{r})|^2\right]\psi(\mathbf{r})=\mu\psi(\mathbf{r}), \quad (8)$$

where the parameter μ is the chemical potential of the species corresponding to the energy required to add a particle in the condensate. The GPE is equivalent, in nonlinear atomic optics, to the nonlinear Schrödinger equation in optics with an external potential term. In addition, the treatment of the mean field in atomic optics is equivalent to the classical treatment of the electromagnetic field in optics. The nonlinear term due to the interaction between particles in the condensate introduces a new energy scale. Thus, in contrast to the linear propagation, new parameter regimes with associated new phenomena and dynamics for special potential parameters are expected [1]. One of the most striking of these is the appearance of solitonic propagation (nonspreading wave packets) and instabilities (i.e., small perturbation of condensate wave function can grow exponentially in time) [1, 7].

2.1.1 External potential

The shape of the external potential $V_{ext}(\mathbf{r})$ depends on the type of trapping. Generally speaking, two types of trapping have been used experimentally, magnetic and optical. In particular, the first BEC was produced using magnetic fields. The first optical confinement of Bose-Einstein condensates (BECs) was achieved in 1998, where the condensate was initially formed in a magnetic trap and subsequently transferred into an optical trap [28]. However, since 2001, it has been demonstrated that BECs can be directly created and confined in optical traps without the need for an intermediate magnetic confinement stage [29]. The main advantage of optical trapping lies in its high degree of flexibility and tunability, enabling precise control over external potentials. A particularly important application of optical traps is the realization of optical lattices, which allow for the simulation of periodic potentials and the exploration of quantum many-body phenomena analogous to those found in solid-state systems [28, 29]. Mathematically, the potential has several forms, depending on the type of trap, such as

- The standard harmonic external magnetic trapping potential

$$V_{ext}(x,y,z)=\frac{1}{2}m\left(\omega_x^2x^2+\omega_y^2y^2+\omega_z^2z^2\right), \quad (9)$$

where the trapping frequencies ω_x, ω_x, and ω_x along the x, y, z directions are generally different.

- The optical lattice potential

$$V_{ext}(x,y,z) = V_L(x,y)\,\sin^2(kz), \qquad V_{ext}(x) = V_0\,\cos^2\left(\frac{2\pi x}{\lambda}\right), \tag{10}$$

where V_L is determined by the transverse intensity profile of the laser beam, k is the wavenumber, and λ is the wavelength of the laser beam.

- The Double well potential

$$V_{ext}(x,y,z) = b\left(x^2 - \frac{a}{2b}\right)^2 + \frac{1}{2}m\omega_\perp^2(y^2 + z^2), \tag{11}$$

where a and b are positive real constants and $\omega_\perp$ is the trapping frequency in the (xy) plane.

- The optical superlattice

$$V_{ext}(z) = V_1\cos(k_1 z) + V_2\cos(k_2 z). \tag{12}$$

This $1D$ stationary superlattice can be formed by sequentially creating two lattice structures using four laser beams. The parameters k_i and V_j represent, respectively, the wavenumbers and amplitudes of the superlattices. Experimental manipulation of these parameters allows precise control of the shape and temporal variation of the external potential.

- Complex time-dependent potential

$$V(x,t) = \alpha(t)x^2 + \lambda(t)x + i\gamma(t), \tag{13}$$

where $\alpha(t)$ is the strength of the harmonic trapping potential, $\lambda(t)x$ can correspond to the gravitational field or to a linear potential, and $\gamma(t)$ is a parameter relating to the gain $\gamma(t) > 0$ or loss $\gamma(t) < 0$ of atoms in the condensate.

- Combination of optical and magnetic potential

$$V_{ext}(x,y,z) = V_0\,\cos^2(kx) + \frac{1}{2}m\left(\omega_x^2x^2 + \omega_y^2y^2 + \omega_z^2z^2\right). \tag{14}$$

- Combination of harmonic and rotating potential

$$V_{ext}(x,y,z) = \frac{1}{2}kr^2 - \omega Lz, \tag{15}$$

where $r^2 = x^2 + y^2 + z^2$, k is the frequency of the harmonic part, ω is the rotation frequency, and $Lz = -i\hbar\left(x\frac{\partial}{\partial y} - y\frac{\partial}{\partial x}\right)$.

It is important to stress that it is now possible to create a wide variety of forms and types (stationary, time-dependent, etc.) of external potential. In the following, we show how the GPE is used to describe soliton in dipolar BECs.

2.2 Soliton in dipolar Bose-Einstein condensates

A soliton is a self-reinforcing, localized wave packet that maintains its shape while it travels at a constant velocity [6–8]. Solitons arise in various physical systems due to a delicate balance between nonlinearity and dispersion (or diffraction). They can occur in fluids, optics, plasmas, Bose-Einstein condensates (BECs), and other areas of physics. Solitons are stable and persist over long distances or times. Even when they interact with other solitons, they tend to re-emerge unchanged, although they might shift in position or phase. These structures are not only of theoretical interest but have practical applications in areas such as telecommunications and quantum mechanics.

Atomic BECs constitute the best environment for studying nonlinear macroscopic excitations such as solitons in quantum systems. For short-range interacting BECs at low temperatures, the wavefunction of the condensate obeys the GPE with cubic local nonlinearity resulting from the interatomic interactions and dispersive effects, which cause the spreading of the wavefunction over time [6, 7]. The one-dimensional (1D) GPE admits stable solitonic solutions for bright (dark) solitons [30, 31] for attractive (repulsive) interactions, equivalent to the self-focusing (self-defocusing) nonlinearity in Kerr media (optical solitons). Dark solitons are characterized by a local density dip and a sharp phase gradient of the wavefunction at the position of the minimum [9]. **Figure 1** shows an example of the density profiles of the bright ($\psi_b(x,0) = \psi_{0b} sech(x)$ as initial condition) and dark ($\psi_d(x,0) = \psi_{0d}\, tanh\,(x)$ as initial condition) solitons as they evolve in time in a simple 1D setting using the GPE framework and the split-step Fourier method.

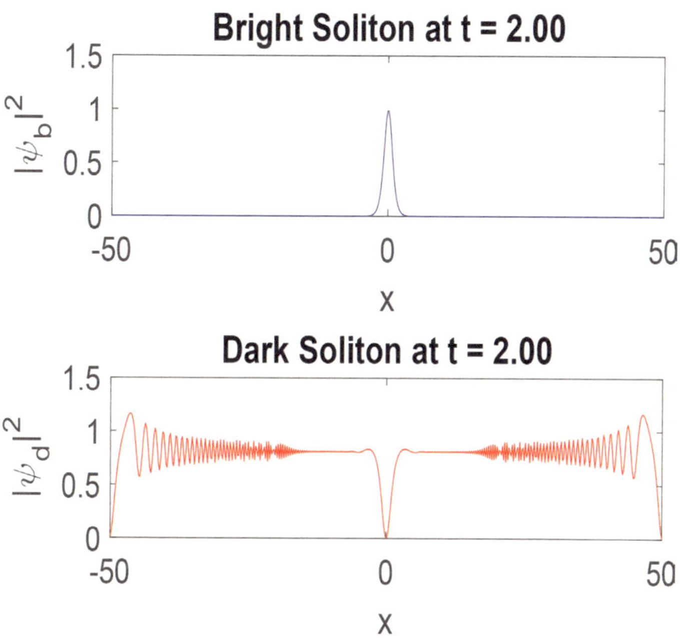

Figure 1.
An example of the density profiles of the bright and dark solitons as they evolve in time.

$$\psi_{0b} = 1 \text{ and } \psi_{0d} = 0.9$$

In BECs, the local nonlinear response is captured by the atom-atom elastic collision. However, at a more detailed level, atoms possess magnetic dipole moments, from the electron and the nuclear spin [17]. The nonlocality arises from the resulting long-range dipole-dipole interaction. From a theoretical point of view, a correction needs to be made to the GPE in order to take nonlocal interaction into account. In other words, the BEC dynamics need a better description of the atom-atom interactions [18]. The dipole-dipole potential interaction between two particles with dipole moments oriented along the unit vectors $\overleftarrow{e}_1$ and $\overleftarrow{e}_2$ is given by

$$V_{dd} = \frac{a_{dd}}{4\pi} \frac{\left(\overleftarrow{e}_1.\overleftarrow{e}_2\right)r^2 - 3\left(\overleftarrow{e}_1.\overleftarrow{r}\right)\left(\overleftarrow{e}_2.\overleftarrow{r}\right)}{r^5}, \tag{16}$$

The constant a_{dd} is a length characterizing the strength of dipolar interaction. When the dipoles are polarized by an external electric or magnetic field in the same direction, meaning that $\overleftarrow{e}_1.\overleftarrow{e}_2 = 1$ and $\overleftarrow{e}_1.\overleftarrow{r} = \overleftarrow{e}_2.\overleftarrow{r} = rcos(\theta)$, the expression of the dipolar interaction becomes

$$V_{dd} = \frac{a_{dd}}{4\pi} \frac{(1 - 3\cos^2(\theta))}{r^3}, \tag{17}$$

θ being the angle between the polarization direction and the relative position of the particles. When the dipolar interaction is taken into account, the GPE takes the form

$$i\hbar \frac{\psi(\mathbf{r},t)}{\partial t} = \left[-\frac{\hbar^2}{2m}\nabla^2 + V_{ext}(\mathbf{r},t) + g|\psi(\mathbf{r},t)|^2 + \int d\mathbf{r}' V_{dd}(\mathbf{r}-\mathbf{r}')|\psi(\mathbf{r}',t)|^2\right]\psi(\mathbf{r},t). \tag{18}$$

In quasi-1D geometry, further simplification can be obtained. This simplification is valid if the BEC is confined in a cylindrically symmetric trap with a transverse trapping frequency $\omega_\perp$ and negligible longitudinal (axis x direction) confinement [18]. In this situation, the dimensionless nonlocal GPE can be obtained by adjusting the new variable $g_d \to a_{dd}\sqrt{l^3\hbar\omega}$ to the local GPE Eq. (7)

$$i\hbar \frac{\psi(x,t)}{\partial t} = \left[-\frac{1}{2}\frac{\partial^2}{\partial x^2} + V_{ext}(x) + g|\psi(x,t)|^2 + g_{dd}\int dx' V_{dd}(x-x')|\psi(x',t)|^2\right]\psi(x,t). \tag{19}$$

Here, ψ is normalized to the number of atoms in the condensate $N = \int_{-\infty}^{+\infty}|\psi|^2 dx$. The space, time, and density are measured in units of the harmonic oscillator length of the transverse trap $a_\perp = \sqrt{\hbar/(m\omega_\perp)}$, $\omega_\perp^{-1}$, and $a_\perp^{-1}$, respectively, where $\hbar$ is the reduced Planck's constant, while $\omega_\perp$ is the frequency of the radial confinement [6, 17]. The scaled parameter $g \equiv 2a_s/a_\perp$ measures the strength of the local binary collisions, where a_s is the scattering length. Thereby, the nonlocality, captured by the parameter g_{dd}, affects notably the formation and the soliton properties. We portray in **Figure 2** the influence of the nonlocal interaction on the density profiles of both bright and dark solitons in a BEC in the framework of Eq. (19).

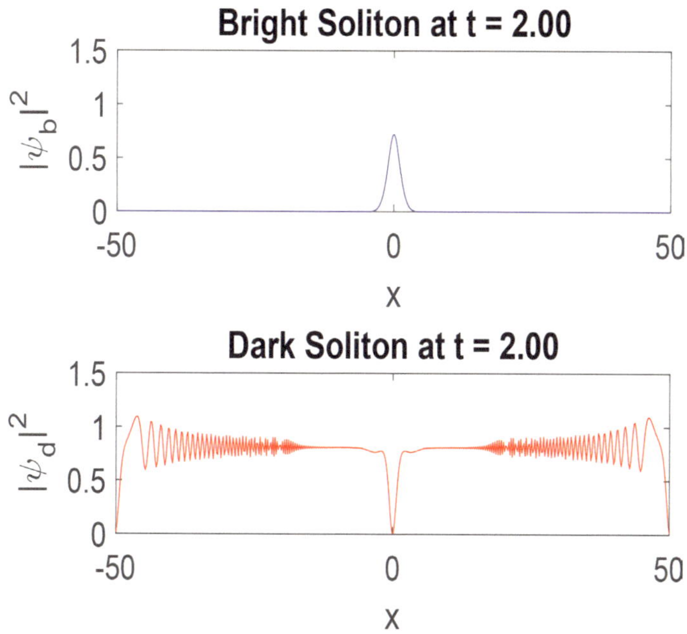

Figure 2.
An example of the density profiles of the bright and dark solitons as they evolve in time in the presence of dipole-dipole interaction (DDI).

2.3 The spin-orbit coupling in Bose-Einstein condensates

Another topic of current interest is the spin-orbit coupling (SOC). In condensed matter systems (such as semiconductors and topological insulators), spin-orbit coupling arises from relativistic effects, where the motion of an electron in an electric field induces a magnetic field that couples to the electron's spin. This coupling plays a crucial role in phenomena like the spin Hall effect and topological insulators.

In ultracold atoms and BECs, while relativistic effects are negligible, synthetic spin-orbit coupling can be engineered using lasers and magnetic fields. This allows physicists to explore the effects of SOC in a highly controlled environment. The spin-orbit coupling in BECs typically involves a two-component condensate where the internal states (often hyperfine states) act as pseudo-spin degrees of freedom. The coupling between these spin states and the motion of atoms in space creates new behaviors that do not occur in regular BECs (without SOC).

Among the different types of SOC, three are of particular interest since they can be implemented in BECs.

- The Rashba SOC takes the form $\lambda\left(\sigma_x p_y - \sigma_y p_x\right)$, σ_x, and σ_y being the Pauli matrices which represent spin components of the atom, while p_x and p_y are the momentum components along the x and y direction, respectively. This coupling is isotropic.

- The Dresslhaus SOC has the form $\lambda\left(\sigma_x p_x - \sigma_y p_y\right)$, and is similar to the Rashba type but has a different symmetry.

- The one-dimensional SOC, where the coupling is between the momentum along one direction (p_x) and the spin along one axis, (σ_y).

Experimentally, it has been demonstrated that the spin-orbit coupling can be achieved synthetically using laser fields. Indeed, two laser beams are used to couple the two internal states of the atom, creating an effective Raman coupling. This coupling changes the atom's internal spin state while simultaneously transferring momentum between the atom and the laser field. The interaction between the spin states and the atom's motion can be described by a synthetic gauge field, leading to the spin-orbit coupling term in the Hamiltonian such as

$$H_{soc} = \frac{p_x^2}{2m} + V(x) + \delta\sigma_z + \frac{\hbar\Omega}{2}\sigma_x, \tag{20}$$

where Ω is the Raman coupling strength, δ is the detuning between the laser fields, and $V(x)$ is the trapping potential. Thus, the SOC introduces new behavior which can modify the excitation spectrum in the dynamics of BEC. Moreover, the presence of SOC changes the dynamics of matter-wave solitons in BECs.

Spin-orbit coupling in Bose-Einstein condensates represents an exciting frontier in cold-atom research. By coupling the internal spin states of atoms to their motion, physicists can simulate and explore complex quantum systems in a controlled, tunable environment. This opens up opportunities for discovering new quantum phases, studying topological phenomena, and realizing novel quantum technologies.

3. The modified Gross-Pitaevskii equation

In this section, we describe the theoretical model that will allow us to analyze the coupled effects of spin-orbit coupling and nonlocal interactions on the dynamics of soliton molecules in a mixture of Bose-Einstein condensates.

3.1 Mean-field description of two-component Bose-Einstein condensates

A trapped two-component Bose-Einstein condensate (BEC) consists of two interacting quantum fluids, which can represent different atomic species, hyperfine states of the same species, or spin components in spinor BECs. The mean-field description of such a system is governed by coupled Gross-Pitaevskii equations (GPEs), which describe the spatial macroscopic wave functions of the two components $\psi = (\psi_1, \psi_2)^T$,

$$\begin{aligned} i\frac{\partial\psi_1}{\partial t} = & -\frac{1}{2}\frac{\partial^2\psi_1}{\partial x^2} - i\delta\frac{\partial\psi_1}{\partial x} + V(x)\psi_1 + g_1|\psi_1|^2\psi_1 + g_{12}|\psi_2|^2\psi_1 + \chi_1|\psi_1|^4\psi_1 \\ & + q\psi_1\int_{-\infty}^{+\infty} R(|x-\xi|)\left(|\psi_1(\xi,t)|^2 + |\psi_2(\xi,t)|^2\right)d\xi + \chi_{12}|\psi_2|^4\psi_1, \end{aligned} \tag{21}$$

$$\begin{aligned} i\frac{\partial\psi_2}{\partial t} = & -\frac{1}{2}\frac{\partial^2\psi_2}{\partial x^2} - i\delta\frac{\partial\psi_2}{\partial x} + V(x)\psi_2 + g_2|\psi_2|^2\psi_2 + g_{12}|\psi_1|^2\psi_2 + \chi_2|\psi_2|^4\psi_2 \\ & + q\psi_2\int_{-\infty}^{+\infty} R(|x-\xi|)\left(|\psi_1(\xi,t)|^2 + |\psi_2(\xi,t)|^2\right)d\xi + \chi_{12}|\psi_1|^4\psi_2. \end{aligned} \tag{22}$$

Here, the parameters $g_{ij} = 4\pi\hbar^2 a_{ij}/m_{ij}$ represent the two-body intraspecies (g_{11}, g_{22}) and interspecies (g_{11}, g_{22}) interactions characterized by the s-wave scattering length (a_{11}, a_{22}) between atoms in the same components and ($a_{12} = a_{21}$) between atoms in the different components [1, 6, 17]. The coefficients χ_1 and χ_2 characterize the intensity of the three-body intraspecies interactions in each component, whereas χ_{12} and χ_{21} account for the three-body interspecies interactions. The wave function ψ_j ($j = 1, 2$) is normalized to the number of atoms in the condensate $N_j = \int_{-\infty}^{+\infty} \left|\psi_j\right|^2 dx$. The real function $V(x)$ describes the external trapping potential. Here, $\delta = \frac{\hbar k_L}{m} p_x$ and k_L are the projected wavenumbers of the Raman beams related to the strength of the SOC [25]. The strength of dipolar interactions is taken into account by the coefficient q.

3.1.1 External optical trap

$V(x)$ models the external optical trap which is a Remoissenet-Peyrard (RP) potential [32, 33].

$$V(x) = v_0 \left[1 + s - \left(1 - s^2\right) \sum_{n=1}^{\infty} s^{n-1} \cos(2nkx)\right], \tag{23}$$

where $-1 < s < 1$ is the deformability of the lattice. The parameter v_0 is the effective depth of the optical potential in units of recoil energy $E_0 = 2\pi\hbar^2/(m\lambda^2)$. $k = 2\pi/\lambda$ is the wavenumber of the OL that can be controlled by varying the angle θ between the two counter-propagating laser beams whose interference creates the optical lattice [28, 29]. The wavelength of the interference pattern is expressed in terms of the angle between the laser beams by the relation $\lambda = (\lambda_{laser}/2)\, sin\, (\theta/2)$, where λ_{laser} is the wavelength of the laser beams producing the optical trap [28, 29]. The Remoissenet-Peyrard (RP) potential is primarily known from the study of solitons and nonlinear dynamics in classical systems, such as nonlinear lattices and condensed matter physics. However, it has been extended and applied to quantum systems, including Bose-Einstein condensates (BECs), where solitons and other exotic states like bright and dark solitons are of interest. The nonlinear Schrödinger equation (NLSE), or Gross-Pitaevskii equation (GPE), is typically used to describe the dynamics of BECs. By incorporating the RP potential into these equations, researchers explore how the nonlinearity of the system interacts with the periodic or quasi-periodic structure of the potential.

3.1.2 Dipolar interaction

The response function $R(x)$ describes the degree of nonlocality [34, 35]. It shows how strongly the properties of the medium at a given location depend on the properties at neighboring regions [34, 35]. We employ a Gaussian response function, which is one of the common models used to describe nonlocal interactions. The Gaussian dipolar interaction in Bose-Einstein condensates (BECs) refers to a specific type of long-range interaction where the dipole-dipole interaction (DDI) is modulated by a Gaussian envelope, often to model experimental setups or modified interactions in ultracold gases. For a BEC with particles having a magnetic dipole moment μ or electric dipole moment d, the dipole-dipole interaction (DDI) is given by Eq. (17). In some cases, the dipolar interaction is modified by a Gaussian envelope to describe finite-range interactions, smooth cutoff effects, or experimental tuning via optical dressing or Rydberg atoms. In this case, the interaction potential is not strictly zero outside a certain range but decays smoothly with distance according to a Gaussian distribution. In this case, the interaction potential between two particles is given by [34, 35]

$$R(x) = \frac{1}{\sqrt{2\pi}\sigma} exp\left(-\frac{x^2}{2\sigma^2}\right), \tag{24}$$

which is normalized to 1, $\int_{-\infty}^{+\infty} R(x)dx = 1$. σ represents the characteristic range of the Gaussian modulation. The exponential decay suppresses long-range interactions, leading to a finite-range dipolar potential. This Gaussian form is particularly useful in numerical simulations to avoid singularities at $x \to 0$ and to mimic realistic experimental setups where interactions are softened by external fields.

3.2 Variational technique

The dynamic evolution of the matter wave associated with the condensate involves several nonlinear phenomena, such as phase self-modulation (intraspecies two-body interactions) and phase inter-modulation (interspecies two-body interactions) due to interatomic interactions. The combined effect of these nonlinear phenomena, dispersion and confinement potential, leads to complex dynamic processes, which are difficult to grasp from the starting nonlinear equation. To make it easier to understand these dynamic processes, we generally try to reduce the dynamics of the soliton, which involves a very large number of degrees of freedom, to that of a simple mechanical system with just a few degrees of freedom. The mathematical techniques used to transform the GPE into a system of ordinary differential equations, describing the evolution of the parameters of the matter wave associated with the condensate, are called *collective coordinate methods*. In this section, we present one of the most common collective coordinate methods, the variational approximation.

3.2.1 Principle of the method

The variational approach was introduced in the cornerstone paper by Anderson [36] to feature solitons in optical fibers, following the pattern of variational approximation for solitons in other physical media such as plasma, which had been developed earlier by Bondeson, Lisak, and Anderson [37]. The variational approach technique for optical solitons was further developed in an important paper by Anderson, Lisak, and Reichel [38]. In 1999, Malomed et al. [39] used the variational approach to study solitons in periodic potentials. These works became the basis for the study of many soliton properties such as existence, stability, and mobility in BECs and nonlinear optics based on variational method [36–39]. To illustrate the principle of the method, we consider a GPE of the form

$$i\frac{\partial\psi}{\partial t} = -\frac{1}{2}\frac{\partial^2\psi}{\partial x^2} + g|\psi|^2\psi + V\psi. \tag{25}$$

This equation can be reformulated as a variational problem corresponding to the Lagrangian density

$$l(\psi) = \frac{1}{2}i\left(\psi^*\frac{\partial\psi}{\partial t} - \psi\frac{\partial\psi^*}{\partial t}\right) - \frac{1}{2}\left|\frac{\partial\psi}{\partial x}\right|^2 - \frac{1}{2}g|\psi|^4 - V|\psi|^2, \tag{26}$$

where ψ^* is the complex conjugate of the wave function ψ. In other words, we can also consider that the GPE results from the variational equations corresponding to the variational principle

$$\delta\int_{-\infty}^{+\infty}\int l dx dt = 0. \tag{27}$$

This principle is explicitly expressed in the form

$$\frac{\partial}{\partial t}\left[\frac{\partial l}{\partial\left(\frac{\partial\psi*}{\partial t}\right)}\right]+\frac{\partial}{\partial x}\left[\frac{\partial l}{\partial\left(\frac{\partial\psi*}{\partial x}\right)}\right]=\frac{\partial l}{\partial\psi*}, \tag{28}$$

leading to Eq. (25).

The aim of the variational approach applied in this case is to replace the nonlinear Gross-Pitaevskii partial differential equation by a system of ordinary first-order differential equations, which is easier to understand. Since the exact solution of the GPE is generally unknown, it is approximated by an ansatz function f given by $f(x,t) = \psi(Q_1, Q_2, \dots, Q_N, t)$, whose expression is known as a function of the variational parameters $Q_1, Q_2, \dots, Q_N$. By inserting the ansatz into the Lagrangian density and integrating according to the spatial variable, we obtain the averaged Lagrangian, defined by

$$L = L\left(Q_j, \dot{Q}_j\right) = \int l dx. \tag{29}$$

The effective Lagrangian leads to a system of variational or Euler-Lagrange equations for the variables Q_j, such that

$$\frac{\partial}{\partial t}\left(\frac{\partial L}{\partial\dot{Q}_j}\right)-\frac{\partial L}{\partial Q_j}=0, \tag{30}$$

which can be solved using analytical or numerical methods.

3.2.2 Implementation of the method

The system of Eqs. (21) and (22) does not possess an analytical solution for arbitrary values of coefficients. Then, to address the problem, we employ the variational approximation, since we aim to analyze the possibility of creating and controlling stable soliton molecules in trapped binary dipolar condensates with the SO coupling in the presence of both intra- and inter-component interactions [36–39]. The Lagrangian density corresponding to the system of coupled equations is

$$l = \sum_{j=1}^{2} l_j - l_{12}, \tag{31}$$

with

$$\begin{aligned} l_j = {} & \frac{i}{2}\left(\psi_j^*\frac{\partial\psi_j}{\partial t}-\psi_j\frac{\partial\psi_j^*}{\partial t}\right)+\frac{i}{2}\gamma\left(\psi_j^*\frac{\partial\psi_j}{\partial x}-\psi_j\frac{\partial\psi_j^*}{\partial x}\right)-\frac{1}{2}\left|\frac{\partial\psi_j}{\partial x}\right|^2-\frac{g_j}{2}\left|\psi_j\right|^4-V(x)\left|\psi_j\right|^2 \\ & -\frac{q}{2}\left|\psi_j\right|^2\int_{-\infty}^{+\infty}R(|x-\xi|)\left(|\psi_1(\xi,t)|^2+|\psi_2(\xi,t)|^2\right)d\xi, \end{aligned} \tag{32}$$

$$l_{12} = g_{12}|\psi_1|^2|\psi_2|^2, \tag{33}$$

where ψ_j^* is the complex conjugate of the wave function ψ_j.

Since the exact solutions of the system (21) and (22) are generally unknown, they are approximated by a trial function or ansatz, the expression of which is known as a function of the variational parameters. In other words, to study solitons and soliton molecules in Bose-Einstein condensates (BECs), especially within the framework of the Gross-Pitaevskii equation (GPE), researchers often use trial solutions to gain insight into the dynamics, stability, and interactions of such nonlinear structures. Below are some common and effective ansatz forms used in this context:

- Sech-shaped ansatz for bright solitons

$$\psi(x,t) = A(t) sech\left(\frac{x - x_0(t)}{a(t)}\right) e^{i(v(t)x+\phi(t))}, \tag{34}$$

where, $A(t)$, $a(t)$, $x_0(t)$, $v(t)$, $\phi(t)$ represent the amplitude, width, center position, velocity, and phase, respectively. This solution is used for an attractive interaction $g > 0$ and describes bright solitons in 1D or quasi-1D BECs.

- Tan-shaped ansatz for dark solitons

$$\psi(x,t) = \sqrt{n_0}\left[\beta(t)\, tanh\left(\frac{x - x_0(t)}{a(t)}\right) + i\alpha(t)\right] e^{-i\mu t}, \tag{35}$$

here, n_0, $a(t)$, $x_0(t)$, $\alpha(t)$, and $\beta(t)$ are the background density, width, center position, soliton depth, and phase jump, respectively. This solution captures the density notch and phase jump in dark solitons. It can be extended to study multiple soliton interactions.

- Two-soliton molecules ansatz

$$\psi(x,t) = A(t)\left[sech\left(\frac{x + d/2}{a(t)}\right) + sech\left(\frac{x - d/2}{a(t)}\right) e^{-i\Delta\phi}\right] e^{-i\mu t}, \tag{36}$$

with $A(t)$, $a(t)$, d, and $\Delta\phi$ being the amplitude, width, separation distance between solitons, and relative phase between the two solitons, respectively. This solution describes a bound state of two interacting bright solitons or a soliton molecule. It helps in exploring binding, repulsion, oscillatory dynamics, and stability.

- Gaussian ansatz

$$\psi(x,t) = A(t)\, exp\left(\frac{(x - x_0(t))^2}{2a(t)^2} + iv(t) + i\phi(t)\right). \tag{37}$$

This ansatz is often used for approximating both bright solitons and multi-component condensates. Suitable for variational methods when soliton profile deviates from ideal sech.

- General form of the Hermite-Gauss ansatz

$$\psi(x,t) = A(t)H_m(x)\,exp\left(\frac{(x - x_0(t))^2}{2a(t)^2} + i\alpha(t)x^2 + i\beta(t) + i\phi(t)\right). \tag{38}$$

The Hermite-Gauss ansatz is especially useful for describing higher-order modes in Bose-Einstein condensates (BECs), soliton trains, and excited states. It is inspired by solutions of the quantum harmonic oscillator and is particularly well-suited for BECs trapped in harmonic potentials, or when studying multi-peak structures like soliton molecules with internal nodes. Here $A(t)$, $a(t)$, $\beta(t)$, $\alpha(t)$, and $\phi(t)$ are variational parameters, standing for the soliton amplitude, width, velocity, chirp, and phase, respectively. $H_m(x) = (-1)^m e^{x^2} d^m e^{-x^2}/dx^m$ is the Hermite polynômial.

We present in this chapter the results obtained by the following first Hermite-Gauss trial function in order to approach a localized bound state in coupled dipolar BECs with spin-orbit coupling, two- and three-body interactions.

$$\psi_1(x,t) = A_1(t)H_1(x)\,exp\left[-\frac{x^2}{2a_1^2(t)} + i\alpha_1(t)x^2 - i\beta_1(t)x + i\phi_1(t)\right], \tag{39}$$

$$\psi_2(x,t) = A_2(t)H_1(x)\,exp\left[-\frac{x^2}{2a_1^2(t)} + i\alpha_2(t)x^2 + i\beta_2(t)x + i\phi_2(t)\right]. \tag{40}$$

We use the variational approach in the symmetric case where $A_1 = A_2 = A$, $a_1 = a_2 = a$, $\alpha_1 = \alpha_2 = \alpha$, $\delta_1 = \delta_2 = \delta$, $g_1 = g_2 = g$, and $\chi_1 = \chi_2 = \chi$. By inserting the ansatz into the Lagrangian density (31) and after integration according to the spatial variable, we obtain the averaged Lagrangian, defined as follows

$$\begin{aligned} L = &-\frac{3\dot{b}A^2a^5\sqrt{\pi}}{2} - \dot{\phi}A^2a^3\sqrt{\pi} + \gamma\beta^2A^2a^3\sqrt{\pi} - \frac{3A^2a\sqrt{\pi}}{4} - \frac{\beta A^2a^3\sqrt{\pi}}{2} - 3\alpha^2A^2a^5\sqrt{\pi} \\ &- \frac{3gA^4a^5\sqrt{2\pi}}{32} - v_0A^2a^3\sqrt{\pi}\left[1 + s - (1 - s^2)\sum_{n=1} s^{(n-1)}(1 - 2n^2k^2)e^{-n^2k^2a^2}\right] \\ &- \frac{3g_{12}A^4a^5\sqrt{2\pi}}{32} + \frac{qA^4a^6\sqrt{\pi}}{16\sqrt{2\pi}}f(a) - \frac{2\chi A^6a^7\sqrt{\pi}}{3\sqrt{3}} - \frac{2\chi_{12}A^6a^7\sqrt{\pi}}{3\sqrt{3}}, \end{aligned} \tag{41}$$

where

$$f(a) = \frac{3a^4 + 4a^2\sigma^2 + 4\sigma^4}{(a^2 + \sigma^2)^{5/2}}. \tag{42}$$

It is important to note that this result has been obtained to study the stability of soliton molecules in coupled dipolar BECs with spin-orbit coupling in the presence of two-body interatomic interactions in the framework of the system (21) and (22) when $\chi_1 = \chi_2 = \chi_{12} = 0$ [31]. In this chapter, we show the influence of the local three-body interactions.

From the Euler-Lagrange Eq. (30), we derive the variational equations for the amplitude, width, and chirp of the soliton molecules as follows

$$\frac{dA}{dt} = -3A\alpha, \tag{43}$$

$$\frac{da}{dt} = 2a\alpha, \tag{44}$$

$$\begin{aligned}\frac{d\alpha}{dt} = \frac{1}{2a^4} - 2\alpha^2 + \frac{gA^2}{16\sqrt{2}} + \frac{g_{12}A^2}{16\sqrt{2}} + \frac{qA^2a^2}{48\sqrt{2}}f'(a) + \frac{6\chi A^4a^2}{\sqrt{3}} + \frac{6\chi_{12}A^4a^2}{\sqrt{3}} \\ - \frac{v_0}{3a}\left(1-s^2\right)\sum_{n=1}\left(2n^2k^2a\right)s^{(n-1)}\left(1-2n^2k^2\right)e^{-n^2k^2a^2},\end{aligned} \tag{45}$$

These equations represent the dynamical evolution of the wave parameters associated with condensates. They allow us to study in detail the properties of soliton molecules in binary dipolar Bose-Einstein condensates with spin-orbit coupling in the presence of local cubic (g, g_{12}) and quintic (χ, χ_{12}) nonlinearities.

4. Soliton molecules

This section is devoted to the presentation of some results. we investigate the dynamical evolution of the two-soliton molecules in the coupled dipolar condensates. The analysis is based on variational and numerical methods.

4.1 Dynamics in the cubic model $\chi_1 = \chi_2 = \chi_{12} = 0$

First, we present the main results obtained in the work of Mboumba et al. [40] where the three-body intra- and interspecies interactions are dropped. We make use of the fourth-order Runge-Kutta method to solve the ODE system above. To analyze the results provided by the variational approach, we take the initial condition $A(0) = 0.29$, $a(0) = 3$, and $b(0) = 0$. The parameters of the optical trap are taken as $v_0 = 10$, $s = 0.3$, and $k = 0.01$.

Figure 3 displays the dynamical evolution of the amplitudes (panel 3(a)) and widths (panel 3(b)) of two solitons forming a molecule in coupled dipolar binary condensates, for various values of the nonlocal interaction strength q. The system includes weak repulsive local intra- and inter-component interactions with $g = 0.01$ and $g_{12} = 0.002$.

Regardless of the value of q, both amplitudes and widths exhibit regular oscillatory behavior around an equilibrium configuration. This indicates the formation of stable matter-wave soliton molecules in the presence of nonlocal dipole-dipole interactions, even when the local interactions are weak and repulsive.

Furthermore, a detailed inspection of the oscillation curves reveals that the frequency of oscillation increases with the strength of the nonlocal interaction q. This behavior suggests that stronger dipolar interactions enhance the vibrational dynamics of the soliton molecule, effectively modulating its internal motion.

Furthermore, **Figure 4(a)** and **(b)** illustrate the time evolution of the amplitudes and widths, respectively, of the binary dipolar condensates for four different values of the nonlocal interaction strength q. In this case, we consider stronger repulsive intra- and inter-component interactions, with $g = 1$ and $g_{12} = 2$.

The results show similar dynamical features to those observed in **Figure 3**, despite the increased local interaction strengths. In particular, the oscillatory behavior of both

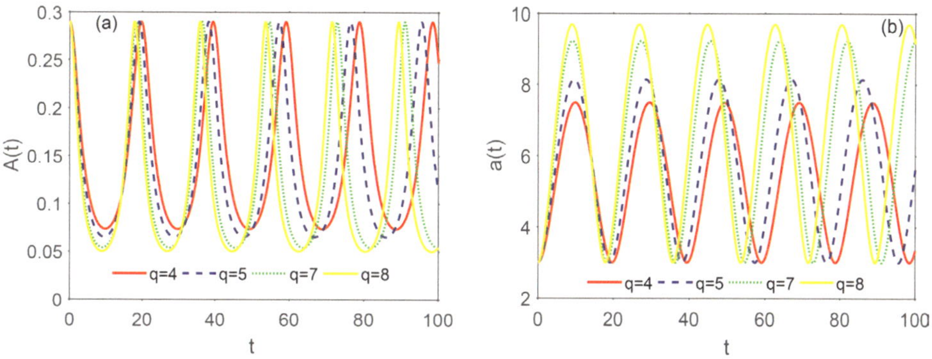

Figure 3.
For four different values of q and $g = 0.01$, $g_{12} = 0.002$, $\chi = 0$, $\chi_{12} = 0$ *in panel (a) The time evolution of the amplitudes of the two-soliton molecules. (b) The time evolution of the widths of the two-soliton molecules. We used* $A(0) = 0.29$, $a(0) = 3$, *and* $b(0) = 0$ *as initial condition.*

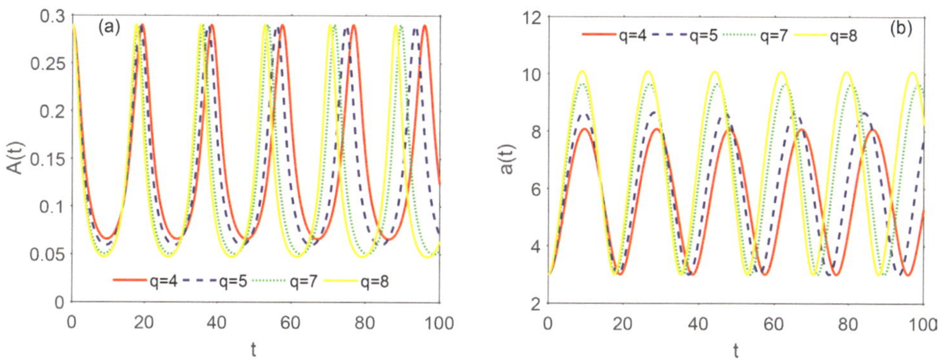

Figure 4.
For four different values of q and $g = 1$, $g_{12} = 2$ *in panel (a) The time evolution of the amplitudes of the two-soliton molecules. (b) The time evolution of the widths of the two-soliton molecules.*

amplitudes and widths persists, indicating that the formation of stable soliton molecules is not restricted to systems with weak local interactions.

These findings confirm that the variational analysis predicts the existence of localized soliton molecules in two-component dipolar Bose-Einstein condensates, even in the presence of strong local binary interactions. The interplay between nonlocal dipole-dipole forces and local repulsion allows for robust molecular binding in both regimes.

To verify the validity of the predictions obtained via the variational approach (VA), we performed a direct numerical integration of the coupled Eqs. (21) and (22). For this purpose, we employed the standard split-step Fourier method. The initial condition used in the simulations corresponds to the ansatz given in Eq. (38). The time evolution was carried out up to $t = 100$, with a sufficiently large spatial grid to eliminate boundary effects and ensure numerical stability.

Figure 5 presents the time evolution of the two-soliton molecules in a mixture of dipolar Bose-Einstein condensates loaded in an optical lattice. The parameters used here are identical to those in **Figure 3**, corresponding to weak local intra- and inter-component interactions ($g = 0.01, g_{12} = 0.002$), with $\delta = 0$ and a strong nonlocal interaction strength $q = 20$.

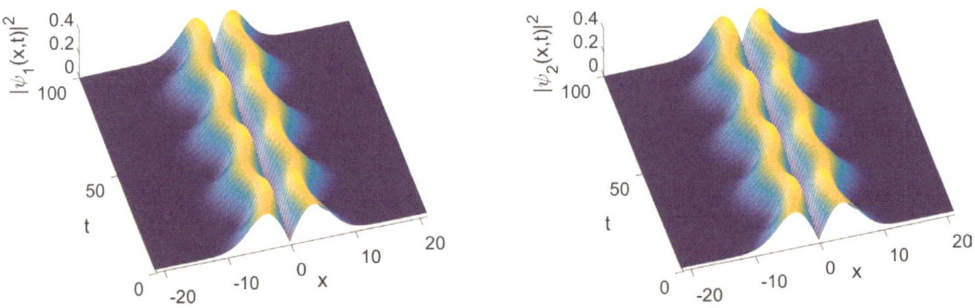

Figure 5.
Evolution of stable soliton molecules in the two-component dipolar BECs with weak intrinsic binary interactions. The parameters are $\sigma = 0.1$, $v_0 = 10$, $s = 0.3$, $k = 0.01$, $g = 0.01$, $g_{12} = 0.002$, $q = 20$, and $\delta = 0$.

The simulations clearly show the stable propagation of soliton molecules in both components of the condensate. The observed oscillatory behavior is attributed to the strong dipole-dipole interactions, which dominate the dynamics and give rise to vibrational motion. These results are in excellent agreement with the predictions of the variational analysis, thereby confirming its accuracy in capturing the essential features of the soliton molecule dynamics in dipolar binary BECs.

In the following, we investigate the combined effects of spin-orbit (SO) coupling and long-range dipole-dipole interactions in the presence of weak binary contact interactions. For this purpose, we introduce a non-zero SO coupling strength $\delta = 0.5$ into the setup of **Figure 5**, and present the corresponding spatio-temporal evolutions of the binary dipolar BECs in **Figure 6**.

The introduction of SO coupling has a pronounced impact on the dynamics of the two-soliton molecules. In particular, we observe a significant distortion in the density profiles of both components. The soliton molecules now exhibit a clear shuttle-like oscillatory motion, characteristic of the flipping-shuttle dynamics induced by SO coupling. This type of behavior has been previously reported in two-component condensates under the combined influence of SO and linear Rabi couplings [41].

Moreover, we observe that the soliton molecules associated with each component interact periodically during propagation, highlighting a rich inter-component coupling mechanism facilitated by the SO term. These flipping-shuttle dynamics may hold potential for practical applications, particularly in the context of information transmission. Due to their localized and robust nature, solitons can travel over long

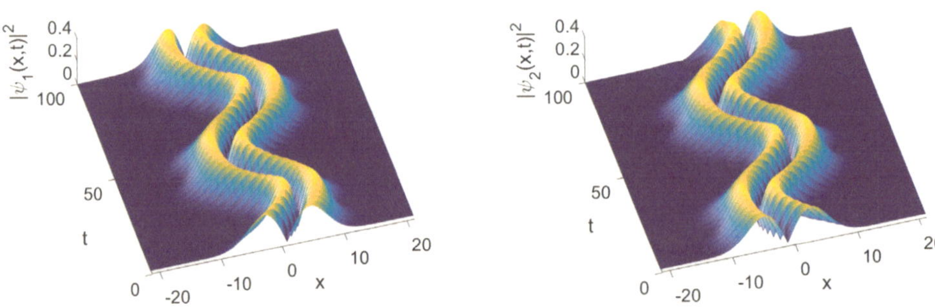

Figure 6.
Impact of the SOC on the density profile of the two-soliton molecules in the two-component dipolar BECs with weak intrinsic binary interactions. The parameters are $\sigma = 0.1$, $v_0 = 10$, $s = 0.3$, $k = 0.01$, $g = 0.01$, $g_{12} = 0.002$, $q = 20$, and $\delta = 0.5$.

distances without deformation, making them promising candidates for communication technologies, such as optical fibers, where signal integrity is crucial.

Their stability means they can propagate through media without distortion, making them valuable in communication technologies, like optical fibers, where maintaining the integrity of a signal over long distances is essential. Solitons often represent a localized concentration of energy that can travel without dissipating. In flipping-shuttle systems, this dynamic might provide insights into how energy can be effectively transported in systems without losing its coherence, which could be applicable in fields like quantum mechanics, material science, or energy transfer technologies. The solitonic behavior could be crucial for controlling or manipulating molecular systems in a precise and stable manner. This could be important in molecular electronics or nanomechanical devices where the dynamics need to be well-controlled.

4.2 Dynamics in the quintic model $\chi_1 = \chi_2 = \chi_{12} \neq 0$

To further explore the nonlinear dynamics of the system, we introduce local three-body interactions in order to investigate the effect of quintic nonlinearities on the soliton molecule dynamics. In **Figure** 7, we present the time evolution of the density profiles of the two-soliton molecules in the binary dipolar BEC mixture, where both the intra- and interspecies three-body interactions are taken to be repulsive.

The results reveal that the solitons comprising the molecule gradually separate from one another during propagation. This observation suggests that repulsive quintic nonlinearities induced by three-body interactions tend to weaken the binding between the solitons, ultimately leading to a repulsive effect.

Therefore, in the presence of spin-orbit coupling, dipole-dipole interactions, and repulsive quintic nonlinearities, the soliton molecules exhibit dispersive behavior. This highlights the significant role that higher-order nonlinear terms can play in destabilizing bound soliton states in complex BEC systems.

In addition, we investigate the influence of attractive three-body interactions on the dynamics of the two-soliton molecules within the system. As shown in **Figure 8**, the distance between the individual solitons comprising each molecule exhibits oscillatory behavior during propagation.

This phenomenon corresponds to a breathing-like dynamic, in which the separation between solitons periodically increases and decreases over time. Such oscillatory motion is a hallmark of nonlinear coupling between the internal degrees of freedom of

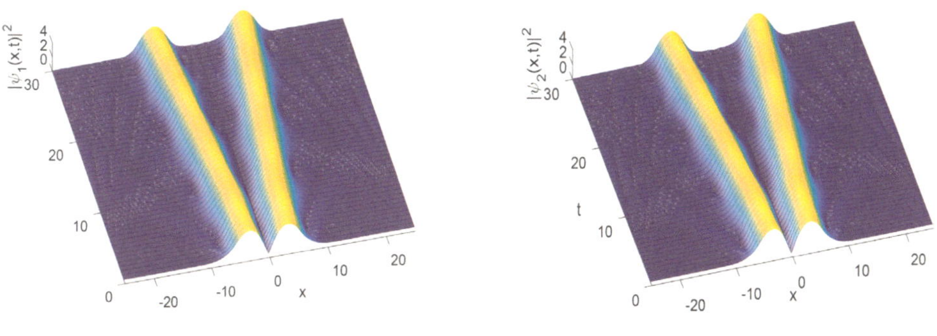

Figure 7.
Impact of the repulsive three-body interactions on the density profile of the two-soliton molecules in the two-component dipolar BECs with weak intrinsic binary interactions. The parameters are $\sigma = 0.1$, $v_o = 10$, $s = 0.3$, $k = 0.01$, $g = -0.03$, $g_{12} = -0.14$, $q = 10$, $\chi = 0.001$, $\chi_{12} = 0.004$, and $\delta = 0.5$.

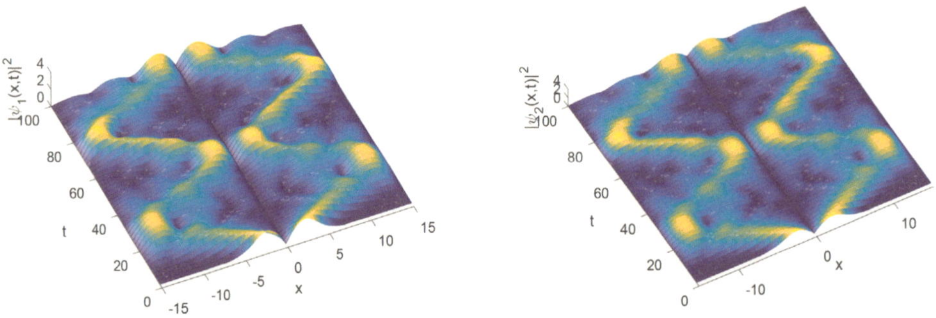

Figure 8.
Impact of the attractive three-body interactions on the density profile of the two-soliton molecules in the two-component dipolar BECs with weak intrinsic binary interactions. The parameters are $\sigma = 0.1$, $v_0 = 10$, $s = 0.3$, $k = 0.01$, $g = -0.03$, $g_{12} = -0.14$, $q = 10$, $\chi = 0.001$, $\chi_{12} = 0.004$, *and* $\delta = 0.5$.

the soliton molecule and the external modulating fields (such as dipolar and spin-orbit interactions).

These results demonstrate that attractive quintic nonlinearities can stabilize bound soliton states through periodic compression and expansion, in contrast to the repulsive case, which leads to dissociation. This breathing behavior may provide useful insights into collective excitations and coherent control of multi-soliton structures in nonlinear quantum fluids.

5. Conclusion

In this chapter, we have demonstrated the controllable formation and complex dynamics of soliton molecules in two-component dipolar Bose-Einstein condensates (BECs) with spin-orbit (SO) coupling. Our approach combined variational methods and direct numerical simulations using the split-step Fourier method, allowing us to explore the interplay between local and nonlocal nonlinearities, including three-body interactions. Following a brief overview of the derivation of the Gross-Pitaevskii equation (GPE) and the motivations for studying dipolar interactions and SO-coupled BECs, we introduced the governing model employed to explore the dynamics of soliton molecules.

Our analysis revealed a broad range of nonlinear dynamical behaviors, such as flipping-shuttle oscillations, breathing modes, and interaction-induced dissociation. These phenomena arise from the intricate balance between dispersion, nonlinearity, and long-range dipolar interactions and are highly sensitive to system parameters. This sensitivity underscores the robustness, tunability, and rich structure of soliton molecules in such hybrid quantum systems.

Beyond practical implications, studying such nonlinear excitations provides valuable insight into the complex dynamics of non-integrable systems, offering a deeper understanding of nonlinear wave phenomena in quantum fluids and contributing to the broader modeling of complex physical, mathematical, and technological systems.

Moreover, beyond their fundamental interest, these soliton structures hold strong promise for technological applications. In particular:

- Their long-lived coherence and internal mode stability make them attractive for quantum information processing, where molecules could act as carriers of qubit-like degrees of freedom or robust elements in quantum memory architectures.

- Their phase sensitivity and nonlinear response to perturbations suggest potential use in precision measurement devices, such as quantum-enhanced interferometers or sensors for weak external fields (e.g., gravitational or magnetic).

- The flipping-shuttle dynamics, which resemble reversible energy or information transport mechanisms, may offer models for biological signal transduction or energy transfer in soft matter systems.

Looking ahead, several avenues for future research emerge:

- While the current study is theoretical, the implementation of such soliton molecules in ultracold atomic setups, especially those with tunable dipolar and spin-orbit interactions, is increasingly feasible with existing optical and magnetic trapping technologies.

- Extending the framework to study soliton lattices or crystalline arrangements of soliton molecules could open pathways to solitonic metamaterials with engineered transport properties.

- Including quantum corrections to the mean-field Gross-Pitaevskii model (e.g., via the Lee-Huang-Yang term or stochastic GPEs) would refine the understanding of stability and coherence in regimes near collapse or fragmentation.

- Investigating how soliton molecules behave in the presence of optical lattices, quasi-periodic potentials, or topological defects could uncover novel localization and transport regimes, relevant for quantum simulation of condensed matter phenomena.

In summary, the study of soliton molecules in SO-coupled dipolar BECs not only enriches our understanding of nonlinear quantum dynamics but also paves the way for innovative applications in quantum technologies, precision metrology, and complex systems modeling. These findings encourage further theoretical and experimental efforts to explore the full potential of soliton-based structures in low-temperature quantum fluids.

Additional information

PACS: 03.75.Lm, 03.75.Kk, 05.45.Yv, 34.20.Cf.

Author details

Maïk Delon Mboumba*, Thierry Blanchard Ekogo and Alain-Brice Moubissi
Laboratoire d'optique Laser et Applications, Département de Physique, Université des Sciences et Techniques de Masuku, Franceville, Gabon

*Address all correspondence to: m.d.mboumba@gmail.com

References

[1] Dalfovo F, Giorgini S, Pitaevskii LP, Stringari S. Theory of Bose-Einstein condensation in trapped gases. Reviews of Modern Physics. 1999;**71**:463

[2] London F. The λ-phenomenon of liquid helium and the Bose-Einstein degeneracy. Nature. 1938;**141**:643-644

[3] Bose SN. Planck's law and light quantum hypothesis. Zeitschrift für Physik. 1924;**26**:178

[4] Anderson MH, Ensher JR, Matthews MR, Wieman CE, Cornell EA. Observation of Bose-Einstein condensation in dilute atomic vapor. Science. 1995;**269**:198

[5] Davis KB, Mewes M-O, Andrews MR, Van Druten NJ, Durfee DS, Kurn DM, et al. Ketterle Bose-Einstein condensation in a gas of sodium atoms. Physical Review Letters. 1995;**75**:3969

[6] Kevrekidis PG, Frantzeskakis DJ, Carretero-Gonzales R. Emergent Nonlinear Phenomena in Bose-Einstein Condensates. Berlin, Heidelberg: Springer; 2008

[7] Kivshar YS, Agrawal GP. Optical Solitons: From Fibers to Photonic Crystals. San Diego: Academic Press; 2003. DOI: 10.1016/B978-012410590-4/50012-7

[8] Malomed BA. Soliton Management in Periodic Systems. New York, NY: Springer; 2019

[9] Burger S, Bongs K, Dettmer S, Ertmer W, Sengstock K, Sanpera A, et al. Dark Solitons in Bose-Einstein Condensates. Physical Review Letters. 1999;**83**:5198

[10] Denschlag J, Simsarian JE, Feder DL, Clark CW, Collins LA, Cubizolles J, et al. Generating solitons by phase engineering of a Bose-Einstein condensate. Science. 2000;**287**:97

[11] Anderson BP, Haljan PC, Regal CA, Feder DL, Collins LA, Clark CW, et al. Watching dark solitons decay into vortex rings in a Bose-Einstein condensate. Physical Review Letters. 2001;**86**:2926

[12] Khaykovich L, Schreck F, Ferrari G, Bourdel T, Cubizolles J, Carr LD, et al. Formation of a matter-wave bright soliton. Science. 2002;**296**:1290

[13] Strecker KE, Partridge GB, Truscott AG, Hulet RG. Formation and propagation of matter-wave soliton trains. Nature (London). 2002;**417**:150

[14] Eiermann B, Anker T, Albiez M, Taglieber M, Treutlein P, Marzlin K-P, et al. Bright Bose-Einstein gap solitons of atoms with repulsive interaction. Physical Review Letters. 2004;**92**:230401

[15] Gross EP. Hydrodynamics of a superfluid condensate. Journal of Mathematical Physics. 1963;**4**:195-207

[16] Pitaevskii LP. Vortex lines in an imperfect Bose gas. J. E. T. P. 1963;**13**: 451-454

[17] Pitaevskii L, Stringari S. Bose-Einstein condensation. Oxford University Press; 2003

[18] Lahaye T, Menotti C, Santos L, Lewenstein M, Pfau T. The physics of dipolar bosonic quantum gases. Reports on Progress in Physics. 2009;**72**: 126401

[19] Griesmaier A, Werner J, Hensler S, Stuhler J, Pfau T. Bose-Einstein condensation of chromium. Physical Review Letters. 2011;**94**:160401

[20] Lu M, Burdick NQ, Youn SH, Lev BL. Strongly dipolar Bose-Einstein condensate of dysprosium. Physical Review Letters. 2011;**107**:190401

[21] Aikawa K, Frisch A, Mark M, Baier S, Rietzler A, Grimm R, et al. Bose-Einstein condensation of erbium. Physical Review Letters. 2012;**108**:210401

[22] Pedri P, Santos L. Two-dimensional bright solitons in dipolar Bose-Einstein condensates. Physical Review Letters. 2005;**95**:200404

[23] Tikhonenkov I, Malomed BA, Vardi A. Vortex solitons in dipolar Bose-Einstein condensates. Physical Review A. 2008;**78**:043614

[24] Sinha S, Santos L. Cold dipolar gases in quasi-one-dimensional geometries. Physical Review Letters. 2007;**99**: 140406

[25] Zhai H. Degenerate quantum gases with spin–orbit coupling: A review. Reports on Progress in Physics. 2015;**78**: 026001

[26] Lin Y-J, Jiménez-Garcia K, Spielman IB. Spin–orbit coupled Bose–Einstein condensates. Nature (London). 2011;**471**:83

[27] Galitski V, Spielman IB. Spin–orbit coupling in quantum gases. Nature (London). 2013;**494**:49

[28] Stamper-Kurn DM et al. Reversible formation of a Bose–Einstein condensate. Physical Review Letters. 1998;**80**(**10**):2027-2030

[29] Barrett MD, Sauer JA, Chapman MS. All-optical formation of an atomic Bose–Einstein condensate. Physical Review Letters. 2001;**87**(**1**):010404

[30] Shun-Jin W, Cheng-Long J, Zhao D, Hong-Gang, Jun-Hong A. Dark and bright solitons in a quasi-one-dimensional Bose-Einstein condensate. Physical Review A. 2003;**68**:015601

[31] Rajendran S, Muruganandam P, Lakshmanan M. Bright and dark solitons in a quasi 1D Bose-Einstein condensates modelled by 1D Gross-Pitaevskii Equation with time-dependent parameters. Phyica D. 2010;**239**:366-386

[32] Remoissenet M, Peyrard M. A new simple model of a kink bearing Hamiltonian. Journal of Physics C. 1981; **14**:L481

[33] Peyrard M, Remoissenet M. Solitonlike excitations in a one-dimensional atomic chain with a nonlinear deformable substrate potential. Physical Review B. 1982;**26**:2886

[34] Turmanov BK, Baizakov BB, Umarov BA, Abdullaev FK. Vibration spectrum of a two-soliton molecule in dipolar Bose–Einstein condensates. Physics Letters A. 2015;**379**:1828-1832

[35] Umarov BA, Aklan NAB, Baizakov BB, Abdullaev FK. Scattering of a two-soliton molecule by Gaussian potential barriers and wells. Journal of Physics Conference Series. 2016;**697**:012023

[36] Anderson D. Variational approach to nonlinear pulse propagation in optical fibers. Physical Review A. 1983;**27**: 3135

[37] Bondeson A, Lisak M, Anderson D. Soliton perturbations: A variational principle for the soliton parameters. Physica Scripta. 1979;**20**:479

[38] Anderson D, Lisak M, Reichel T. Asymptotic propagation properties of pulses in a soliton-based optical-fiber communication system. Journal of the Optical Society of America B: Optical Physics. 1988;**5**:207

[39] Malomed BA, Wang ZH, Chu PL, Peng GD. Multichannel switchable system for spatial solitons. Journal of the Optical Society of America B: Optical Physics. 1999;**16**:1197

[40] Mboumba MD, Makoundit GJN, Sadem CK, Moubissi AB, Kofané TC. Soliton molecules in coupled dipolar Bose–Einstein condensates with spin-orbit coupling. Modern Physics Letters B. 2023;**37**:2350075

[41] Sakaguchi H, Malomed BA. Flipping-shuttle oscillations of bright one- and two- dimensional solitons in spin-orbit-coupled Bose-Einstein condensates with Rabi mixing. Physical Review A. 2017;**96**:043620

Chapter 5

Motion Control of Suspended Loads via Magnetic Interaction

Panagiota Atzampou, Peter C. Meijers, Apostolos Tsouvalas and Andrei V. Metrikine

Abstract

This chapter presents a coupled contactless control strategy designed to simultaneously manage the translational and torsional motions of a suspended load. The approach relies on magnetic dipole interactions between two pairs of magnets. Translational motion is controlled by modulating the polarity and strength of an external dipole, while torsional motion is regulated by adjusting the orientation of the external magnetic field of the corresponding controller dipole. The findings show that the system effectively dissipates motion caused by external disturbances and non-trivial initial conditions. Crucial control factors include the separation distance between the magnetic pairs and the capacity of the translational control to counterbalance the attractive forces induced by torsional torque. This magnetic control technique offers a strong basis for implementing non-contact position control of large payloads such as offshore wind turbine components.

Keywords: magnetic control, magnetic stiffness, magnetically induced damping, magnetic dipole, PID control, active vibration control, offshore wind turbine installation

1. Introduction

Controlling the motion of suspended loads is critical for maintaining safety and operational efficiency in offshore wind turbine installations. In deep-sea environments, these installations are often performed using floating vessels (**Figure 1**). While such vessels provide impressive lifting capacity, they are significantly affected by wave-induced motion, which propagates to the crane and payload system. A variety of motion compensation strategies have been introduced over time, including active tugger systems [2], gripper frames [3], and crane-based compensators [4, 5]. Nonetheless, these approaches usually involve mechanical attachments and manual intervention. Considering the high-precision demands and limited weather windows during installation procedures, there is a notable research gap regarding non-contact motion compensation technologies tailored for such applications [6, 7].

The technique introduced relies on magnet-to-magnet interaction between the payload and a set of electromagnetic actuators, utilising both attractive and repulsive forces. To validate the concept, a scaled pendulum-based model is developed and

Figure 1.
Semi-submersible crane vessel (HMC-Sleipnir) performing a floating offshore installation of an OWT tower [1].

examined, focusing on the controllability of a two-degree-of-freedom dynamic system. This system exhibits vibrational behaviour in both translational and torsional modes, which play a critical role in determining assembly precision. The control strategy aims to mitigate the influence of a predefined pivot motion and initial conditions, ensuring the payload remains at a fixed desired target position.

This chapter first introduces the dynamical system, emphasising the magnetic coupling. Then, the control design is presented with a focus on stability requirements. Finally, the controlled response of the system is demonstrated for selected initial conditions and external disturbances. Special attention is given to the efficiency of the combined control and its impact on the system's dynamical behaviour, including the nonlinear coupling effects from magnetic interactions.

2. Methodology

2.1 Governing equations of a controlled planar torsional magnetic pendulum

A lab-scale simple pendulum analogue with length ℓ is augmented with a cylindrical mass M that is free to rotate around its shaft (longitudinal axis). The coupled model is illustrated in **Figure 2**, while the dimensions and properties of the system are given in **Table 1**. In principle, the set-up combines two SDOF systems reported in previous work [6, 7]. The control design requires the action of two separate electromagnetic actuators (EMs): one controlling the rotation of the mass in xz-plane, ψ (translation) and the other in xy-plane, ϕ (torsional rotation). In the envisioned

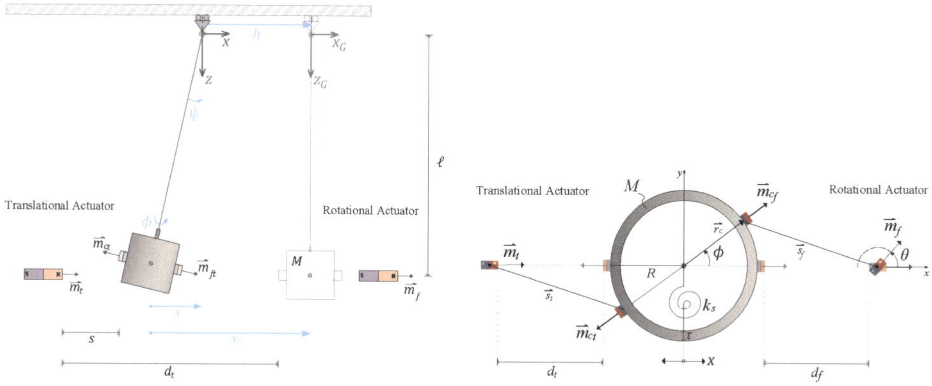

(a) xz-plane (b) xy-plane

Figure 2.
Schematic of the pendulum set-up for the combined control of its translational and torsional degree of freedom.

M [kg]	ℓ [m]	R [mm]	τ [mm]	k_s [N mm/°]	I_z [kg mm^2]	f_n [Hz]
0.2	1.0	20.0	2.0	50	36.2	[0.5, 4.2]

Table 1.
Set-up parameters of the dynamic 2DOF system.

practical application, the suspended mass is an elongated cylindrical structure (e.g., an OWT monopile foundation). Such geometry allows for an advantageous placement of contactless control points at different heights along its length on one side of the load, ideally on the floating vessel's deck. For simplicity, in this study of the combined control, two permanent magnets (PMs) are attached to opposite sides of the cylindrical mass, represented by magnetic dipoles.

Potential interactions between the magnetic fields of the mounted dipoles or direct interference from one actuator affecting the field of the other, as experienced by the targeted fixed magnet, are neglected in the current analysis. Such interactions would not occur in practical applications, due to adequate separation distance between the control points and respective actuators (along the structure's z-axis). Note that the suspended load is assumed to be adequately small to be considered as a point mass in the derivations. The motion of the pivot is described by $h = A_h \sin(\Omega t)$.

Two electromagnetic actuators are placed at a fixed distance on either side of the load, d_t for the translational and d_f for the torsional controller. I_z is the moment of inertia of the mass around the z-axis, and k_s is analogous to the stiffness provided by the suspension cable and experienced by the mass rotating in this direction (**Figure 2b**).

The equations of motion of this two-degree system are the following [8, 9]:

$$M\ell^2\ddot{\psi} + Mg\ell\psi = -M\ddot{h}\ell + T_t, \tag{1}$$

$$I_z\ddot{\phi} + k_s\phi = T_f, \tag{2}$$

where the external magnetic torques are equal to

$$T_t = -\frac{\partial W_m}{\partial \psi}, \tag{3}$$

$$T_f = -\frac{\partial W_m}{\partial \phi}. \tag{4}$$

In the above, W_m is the contribution of the pairs of interacting magnetic dipoles to the potential energy of the system and is equal to

$$W_m = -\sum \vec{m}_{cn} \cdot \vec{B}_n, \tag{5}$$

in which $\vec{B}_n$ is the magnetic field exerted by the EM ($\vec{m}_n$) to the location of the respective PM on the cylindrical point mass ($\vec{m}_{cn}$), where the general subscript notation n is either t for the translational or f for the torsional control. The magnetic field $\vec{B}_n$ reads

$$\vec{B}_n = -\frac{\mu_0}{4\pi}\vec{\nabla}\frac{\vec{m}_n \cdot \vec{s}_n}{s_n^3}, \tag{6}$$

in which $\vec{s}_n$ denotes the vector representing the separation distance between the two dipoles (see **Figure 2b**). The magnitude of s_n is calculated by the following norm:

$$s_n = \|\vec{P}_n - \vec{P}_{cn}\|, \tag{7}$$

with the coordinates of the magnetic dipoles given by

$$\vec{P}_n = \begin{bmatrix} i_n(R + d_n) \\ 0 \\ \ell \end{bmatrix} \text{ and } \vec{P}_{cn} = \begin{bmatrix} \ell\sin(\psi) + h \\ 0 \\ \ell\cos(\psi) \end{bmatrix} + i_n[R\sin(\phi)R\cos(\phi)0], \tag{8}$$

with $i_n = -1$ or $+1$ for the translational and the rotational control, respectively.

Referring to **Figure 2b**, the vectors representing the four magnetic dipole moments are given by

$$\vec{m}_t = M_t\begin{bmatrix} 1 \\ 0 \\ 0 \end{bmatrix} \text{ and } \vec{m}_{ct} = -M_{ct}[\cos(\phi)\sin(\phi)0] \tag{9}$$

for the translational control, and

$$\vec{m}_f = M_f[\cos(\theta)\sin(\theta)0] \text{ and } \vec{m}_{cf} = M_{cf}[\cos(\phi)\sin(\phi)0] \tag{10}$$

for the torsional control, where θ is the orientation angle of the torsional actuator's magnetic field.

After substitution of the above quantities into Eq. (5), the resulting magnetic potential energy W_m is a nonlinear trigonometric expression. In order to analyse the dynamics of the coupling, the trigonometric terms of ϕ and ψ are replaced by their corresponding Taylor series expansions up to the second order around the equilibrium points $\phi = \psi = 0^{\circ}$. The contribution from the magnetic interaction can then be concisely rewritten as:

$$W_m = -K_{m_1}\psi + K_{m_2}\phi + K_{m_3}\psi^2 + K_{m_4}\phi^2 + K_{m_5}\psi\phi + C, \tag{11}$$

with the coefficients:

$$K_{m_1} = 3\ell\left(\frac{M_{ct}M_t\mu_0}{2\pi(d_t+h)^4} + \frac{M_{cf}M_f\mu_0}{2\pi(d_f-h)^4}\cos(\theta)\right), \tag{12}$$

$$K_{m_2} = \frac{M_{cf}M_f\mu_0}{4\pi(d_f-h)^4}(3R+d_f-h)\sin(\theta), \tag{13}$$

$$K_{m_3} = 3\ell^2\left(\frac{M_{ct}M_t\mu_0}{\pi(d_t+h)^5} - \frac{M_{cf}M_f\mu_0}{\pi(d_f-h)^5}\cos(\theta)\right), \tag{14}$$

$$\begin{aligned} K_{m_4} = &\frac{M_{cf}M_f\mu_0}{4\pi(d_f-h)^5}\left((d_f-h)^2 + 6R^2 + 6Rd_f - 6Rh\right)\cos(\theta) \\ &- \frac{M_{ct}M_t\mu_0}{4\pi(d_t+h)^4}(3R+d_t+h), \end{aligned} \tag{15}$$

$$K_{m_5} = 3\ell\left(\frac{M_{cf}M_f\mu_0}{4\pi(d_f-h)^5}(4R+d_f-h)\sin(\theta)\right), \tag{16}$$

$$C = -\frac{M_{cf}M_f\mu_0}{2\pi(d_f-h)^3}\cos(\theta) + \frac{M_{ct}M_t\mu_0}{2\pi(d_t+h)^3}. \tag{17}$$

The magnetic interaction couples the two degrees of freedom ψ and ϕ, through the coupling term K_{m_5} in the magnetic potential energy contribution in Eq. (11), specifically, the external torques introduce magnetically induced stiffness-like terms (proportional to state variables). The coupling is exclusively attributed to torsion, as K_{m_5} contains important parameters of the torsional control. Note that θ is not an unknown variable, but rather a control parameter.

After substitution of the relations above, the magnetic torques in Eqs. (1) and (2) are given by

$$T_t = +K_{m_1} - 2K_{m_3}\psi - K_{m_5}\phi, \tag{18}$$

$$T_f = -K_{m_2} - 2K_{m_4}\phi - K_{m_5}\psi. \tag{19}$$

Therefore, the equations of motion that govern the system can be reformulated into matrix form as follows:

$$\mathbf{M}\ddot{\vec{q}} + \mathbf{K}\vec{q} = \vec{T}, \tag{20}$$

$$\begin{bmatrix} M\ell^2 & 0 \\ 0 & I_z \end{bmatrix}\begin{bmatrix} \ddot{\psi} \\ \ddot{\phi} \end{bmatrix} + \begin{bmatrix} Mg\ell + 2K_{m_3} & K_{m_5} \\ K_{m_5} & k_s + 2K_{m_4} \end{bmatrix}\begin{bmatrix} \psi \\ \phi \end{bmatrix} = -\begin{bmatrix} M\ell\ddot{h} - K_{m_1} \\ K_{m_2} \end{bmatrix}, \tag{21}$$

in which the bold characters **M** and **K** denote the mass and stiffness matrices, respectively. The coupling present in the off-diagonal elements of the stiffness matrix is attributed exclusively to torsion, as K_{m_5} depends only on terms which are influenced by the torsional controller.

2.2 Control strategy

The translational vibrations are controlled by adjusting the strength and polarity of the dipole M_t [6]. At the same time, the orientation of the magnetic field of M_f is used to control the torsional rotation of the cylinder. In Eqs. (18) and (19), the terms K_{m_1} and K_{m_2} serve as the controlled external forces acting on the system. The magnetic strength of the translational controller is governed by a proportional-derivative (PD) control equation, expressed as

$$M_t = K_p e + K_d \dot{e}, \text{with } e = \xi - \ell\psi - h, \tag{22}$$

where the control gains are constant values with $K_p, K_d > 0$, e is the error based on the translation x_n of the moving load inline with the magnetic dipole pair, and ξ represents the desired translational motion. The control rule for the torsional mode is defined by

$$\theta = \Delta\theta \tanh(\beta\dot{\phi}), \text{with } \Delta\theta = \text{const.}, \tag{23}$$

where $\Delta\theta$ is the fluctuation step for the orientation of the magnetic actuator θ. Moreover, the parameter β controls the steepness of the transition between the values around zero. As β approaches infinity, the control function converges to the standard sgn(..) function, while still providing a smooth and differentiable approximation.

Figure 3 presents the torsional damping control strategy within the phase-space quadrants, illustrating how energy is dissipated through the control action. A small

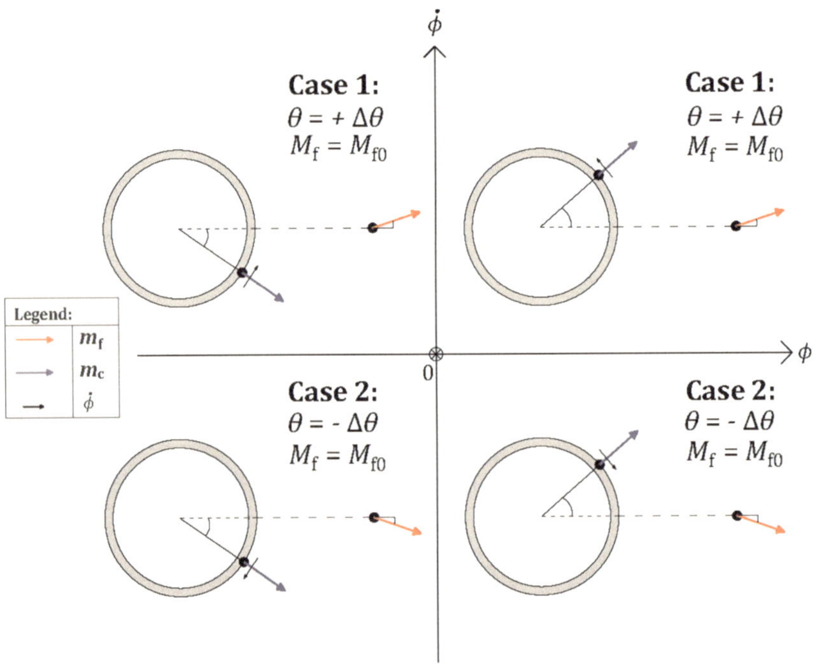

Figure 3.
Phase portrait of the torsional damping control scheme detailing the control cases per quadrant [7].

fluctuation of the controller's dipole orientation is employed to produce a torque counteracting the system's motion, effectively mimicking a Coulomb friction force. At each time step, the control algorithm evaluates a set of predefined scenarios based on the deviation e_f from the reference position (static equilibrium at $\phi = 0^\circ$).

The control cases are:

- Case 0: No action taken ($M_f, \theta = 0^\circ$), when the damping control is successful and the attenuated response is within negligible amplitude bounds ($e_f < 0.5^\circ$ and $\dot{e}_f < 50^\circ/s$), and
- Case 1: Positive EM rotation ($M_f, \theta = +\Delta\theta$), or
- Case 2: Negative EM rotation ($M_f, \theta = -\Delta\theta$).

2.3 Stability requirements

While the signs of the dipoles of the torsional control and the mounted dipole M_{ct} are known and taken into account in the derivations, M_t is subject to an alternating polarity dictated by the error (e) and the error's rate of change ($\dot{e}$). Moreover, the strength is a controlled variable and, as numerically described in Eq. (22), introduces additional stiffness (through the $K_p\psi$) and damping ($K_d\dot{\psi}$) terms to the system. Therefore, it is difficult to deduce one concise rule for the stability of the system in terms of stiffness due to the presence of time-varying functions in the stiffness coefficient in Eq. (21).

To ensure a stable response, a condition concerning the magnetic strength of the translational actuator must be satisfied. This condition directly affects the stability of the controlled dynamical system. Instability is undesirable as it results in an increasing amplitude of the response, which, within the proposed control scheme, can reduce the separation distance between the magnets. Such a reduction could potentially lead to the "failure" of the control, as one of the pairs of magnetic dipoles may yield to the attracting forces and stick to each other.

To assess the stability of the system, the natural frequencies ω_n need to be calculated. These frequencies are determined by solving the eigenvalue problem [10]. The characteristic equation is given as

$$\det\left(\mathbf{K} - \omega_n^2\mathbf{M}\right) = 0, \tag{24}$$

in which $\mathbf{K}$ and $\mathbf{M}$ are the stiffness and mass matrices as defined in Eq. (21). Non-trivial, real, and positive roots of ω_n^2 correspond to the natural frequencies of a stable system. However, if at least one ω_n has a positive imaginary part, or equivalently if $\omega_n^2 < 0$, the system falls within the unstable regime. For reference, the natural frequencies of the uncontrolled system, derived from the modal analysis, are equal to: $\omega_n = [3.13, 26.38]\mathrm{rad/s}$, which correspond approximately to $f_n \approx [0.50, 4.20]$ Hz.

For a distance ratio $\kappa = d_t/d_f = 1$, namely equal separation distances, a desired equilibrium $\phi = 0^\circ$, and equal magnetic dipole strengths $M_f = M_{cf} = M_{ct}$, the natural frequencies ω_n^2 are plotted against different magnetic strength ratios $\lambda = M_t/M_f$ in **Figure 4**. The fluctuation angle $\Delta\theta$ is an *a priori* selected small angle (as per the control rule presented in Eq. (23)). Thus, it is reasonable to assume the small angle approximation rule applies, namely $\cos(\theta) \approx 1$ and $\sin(\theta) \approx \theta$. Now, the only time-varying variable of the system is the pivot point motion. Therefore, it is important to examine the changes in the natural frequencies derived for three distinct values of h. For $h = 0$, the system remains stable for $\lambda < -1$ (**Figure 4a**). However, the stability range broadens when the pivot moves away from the torsional actuator (**Figure 4c**). Conversely, when the distance between the two translational dipoles increases ($h > 0$, **Figure 4b**), the range shifts toward lower ratios. In order to compensate for the attractive forces of the torsional dipoles and the additional leverage of a smaller d_f, the translational actuator must increase its (absolute) strength by over threefold to maintain a stable response.

The common denominator in **Figure 4** is the fact the ratio remains negative across all three amplitudes of suspension motion h. This is a logical condition for the stability of the system, since when M_t is positive, the dipoles have opposite polarities, resulting in repulsion.

Figure 5a illustrates the normalised magnetic potential (relative to its maximum) as a function of the angle ϕ for various orientation angles θ, under translational control with repulsive interaction. This configuration corresponds to a potentially unstable state for the pendulum. In contrast, when the controller's dipole strength is negative, the dipoles align to generate attractive forces on both sides, resulting in a stable

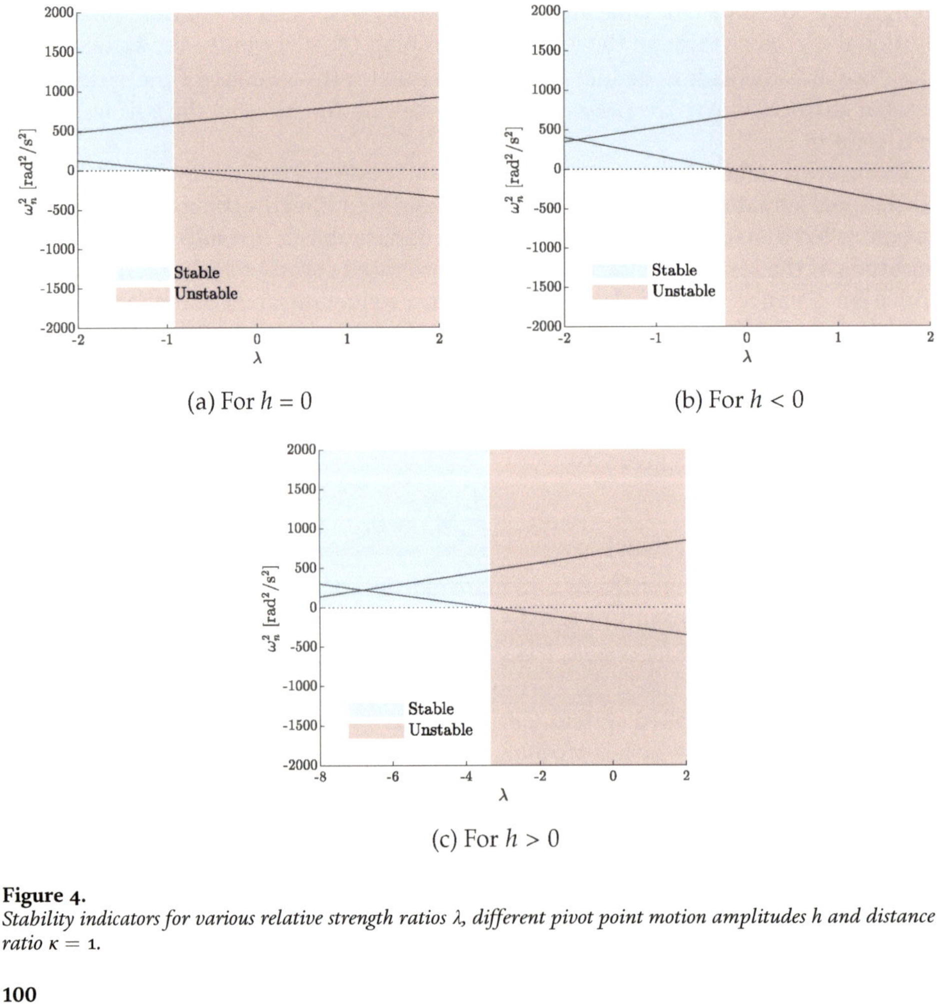

(a) For $h = 0$

(b) For $h < 0$

(c) For $h > 0$

Figure 4.
Stability indicators for various relative strength ratios λ, different pivot point motion amplitudes h and distance ratio $\kappa = 1$.

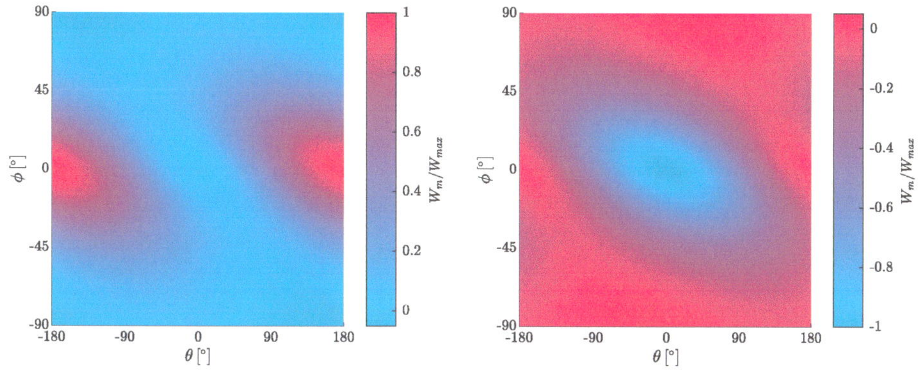

(a) Repulsion: $M_t M_{ct} < 0$ (b) Attraction: $M_t M_{ct} > 0$

Figure 5.
Normalised magnetic potential over its maximum value W_m/W_{max} against ϕ for various angles θ and $\psi = h = 0$. The magnetic dipoles are equal to $M_f = -M_{ct} = M_{cf} > 0$, with equal initial separation distances, $\kappa = 1$.

configuration. This is characterised by a single global potential well, as shown in **Figure 5b**.

Inspecting K_{m_1} as given by Eq. (12), two distinct torque actions are observed: one direct, exerted by the translational controller, and another indirect, generated by the horizontal components of the attractive force between the magnetic dipoles of the torsional controller. For an effective controlled response, the translational controller should counteract the additional torque around the pivot point induced by the torsional control. It is important to note that the translational control torque does not directly influence the motion in the xy-plane, except for the excitation amplitude at the pivot, as outlined in Eq. (13).

Therefore, it is suspected that the time trace of M_t will have a non-zero average value M_{t_0} around which the translational control variable will fluctuate with respect to the error and suspension motion. The amplitude of M_{t_0} arises from the main prerequisite for the stability of the system, neutralising the torque generated by M_f in Eq. (12), which yields

$$K_{m_1} = 0 \rightarrow \lambda = \frac{M_{t_0}}{M_f} \approx -\frac{(d_t + h)^4}{(d_f - h)^4}. \tag{25}$$

Figure 6a depicts the condition K_{m_1} for various strength ratios λ, with a distance ratio $\kappa = 1$, across a range of pivot excitation amplitudes. The strength ratios λ identified through stability analysis (**Figure 4**) are validated and correspond to the highlighted points in the figure.

For a fixed strength ratio of $\lambda = -1$, the necessary distance ratios κ for achieving a stable controlled response are illustrated in **Figure 6b**. The nonlinear relationship between the distance and the term K_{m_1} is evident. When the pivot reaches its maximum positive displacement, the translational controller must be positioned closer to the mass to maintain its effectiveness.

By comparing the two stability checks in **Figure 6**, the strength ratio λ varies significantly with the pivot point compared to the distance ratio κ. Positive pivot amplitudes h generally introduce greater sensitivity in both parameters, requiring finer tuning for maintaining stability.

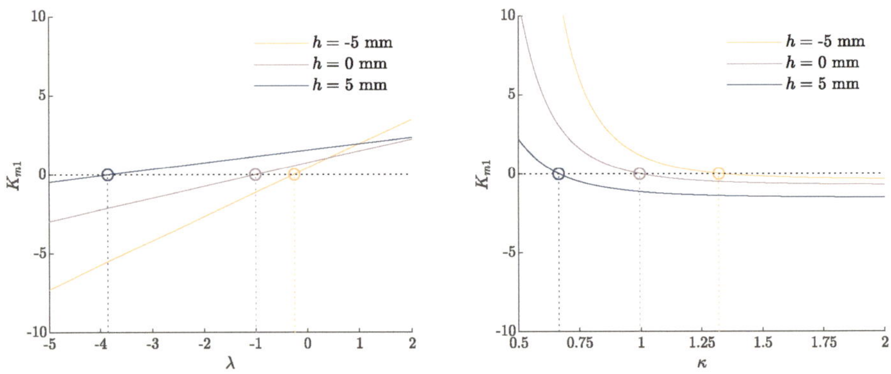

(a) For different λ and $\kappa = 1$ (b) For different κ and $\lambda = -1$

Figure 6.
Stability check $K_{m_1} = 0$ for different pivot point amplitudes.

3. Control of translational and torsional vibrations

The control mode of interest pertains to the motion attenuation of the translational and torsional vibrations while subjected to an external excitation of its pivot and non-trivial initial conditions. The desired position of the suspended load is chosen $\psi = \phi = 0^\circ$. The parameters of the controlled system are presented in **Table 2**.

For translational control, the PD controller's gains are determined heuristically to minimise error and enhance the controller's response. Thus, in this exemplary case, the proportional gain is set as $K_p = 3 \cdot 10^4$ A m and the derivative gain as $K_d = 30$ A m s. The control parameters for the rule defined in Eq. (23) are set as follows: the steepness coefficient β is assigned a value of 1, and the orientation adjustment step is chosen as $\Delta\theta = 5^\circ$. Both controllers are activated at $t_s = 2$ s from the start of the simulation. This delay may lead to a more intricate transient response, aligning with the real application where vibrations would be present before the controllers are activated.

Figure 7 displays the controlled (solid lines), desired (dotted line) and uncontrolled responses of the system (dashed line). More specifically, **Figure 7a** demonstrates the effectiveness of the PD translational controller, achieving minimal steady-state error and a quick, smooth transient response, as indicated by the displacement and velocity time series. The different distances affect the strength of the magnetically induced damping, with $d = 30$ mm exhibiting the fastest dissipation. Notably, the control variable M_t, after successful translational control, settles at $M_t = -M_{t_0} = -M_f$, as predicted by Eq. (25).

$M_f M_{c_f}$ [A^{2}m^4]	M_{c_t} [Am^2]	κ [−]	d_t [mm]
5	−1	1	[30, 40, 50]
A_h [mm]	f_h [Hz]	ϕ_0 [°]	$\dot{\phi}_0$ [°/s]
5	0.6	20	0

Table 2.
System parameters for the combined controlled vibrations.

The controlled response of the torsional degree of freedom is depicted in **Figure 7b**. In this case, the attenuation of motion is more gradual than in the translational response, displaying a damping mechanism similar to Coulomb friction, which is evident by the linear decay of the oscillations. To reduce the duration of the decay, the electromagnet strength M_f should be increased. Alternatively, the closer proximity of the two interacting dipoles yields similar favourable results (**Figure 7b**).

On a general note, the different frequencies of the oscillation observed in the steady-state responses of the two degrees of freedom reflect the natural frequencies of the uncontrolled system.

In **Figure 7c, d**, the two control torques exerted on the load are presented. The torque time series resembles the shape of the translational displacement and torsional angle time series. The highest amplitudes are observed for the translational torque T_t, whereas the torsional torque exhibits higher variation in the steady-state. Interestingly, the range of the translational torque is two orders of magnitude greater than that of the torsional torque, highlighting the difference in control demands between the two degrees of freedom.

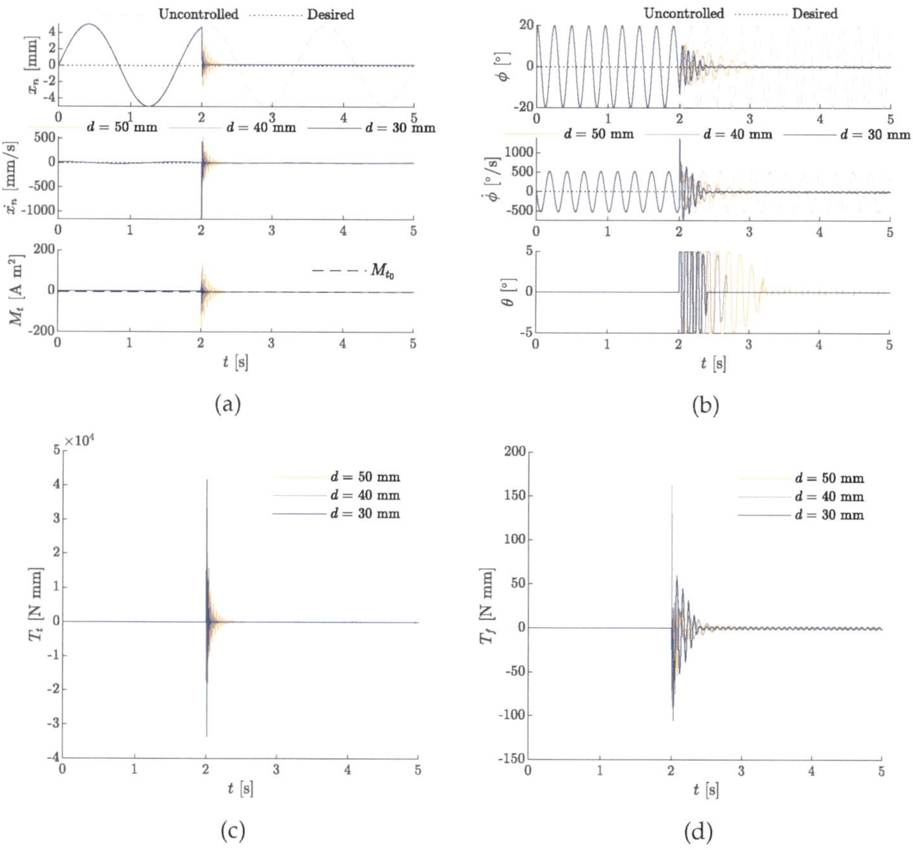

Figure 7.
Combined control efficiency metrics for motion attenuation. (a) Controlled response for translational vibrations. (b) Controlled response for torsional vibrations. (c) Effective actuation torque T_t for the translational control. (d) Effective actuation torque T_f for the torsional control.

Similar trends are observed in the actuation power time series shown in **Figure 7e, f**. The actuator power is computed as

$$P_{t,f} = T_{t,f}\dot{q}, \tag{26}$$

where the angular velocity $\dot{q}$ corresponds to $\dot{\psi}$ or $\dot{\phi}$, depending on the rotational degree of freedom under consideration.

As the actuator distance d decreases, the control effort becomes more pronounced, reflecting stronger magnetic interactions and consequently greater control authority at shorter distances—consistent with the inverse relationship between magnetic force and distance. The difference in scale between the plots (approximately one order of magnitude) validates that the controller exerts significantly more effort for translational control.

An interesting observation emerges from **Figure 7d, f**: the actuator's torque and power reach higher maximum amplitudes at $d = 40$ mm, while more pronounced oscillations occur at $d = 30$ mm. This suggests that reducing the distance may enable quicker damping of torsional oscillations, but it does not necessarily represent the optimal separation distance of the interacting magnets in terms of actuator force.

4. Conclusions

In this chapter, the combined magnetic control of two key degrees of freedom (translation and torsion) is analysed. The coupling presented due to the simultaneous action of the two actuators is reflected in the additional magnetically induced stiffness terms in the equations of motion of the system. The mean amplitude of the translational control is dictated by the chosen amplitude of the torsional actuator dipole that is necessary to counteract the attraction forces exerted by the torsional dipole arrangement on the suspended load.

A combined control approach is evaluated for the mode of disturbance rejection for both translational and torsional motions. The controllers perform effectively, significantly reducing vibrations, with the translational response showing a rapid transition to the target steady-state. The success of the combined control lies primarily in the precise motion control applied by the translational actuator, which determines the separation distance of the torsional dipoles, which is a critical factor in achieving efficient torsional control. The performance of the torsional controller can be further optimised by modifying the separation distance, magnetic strength, and orientation step of the external magnetic dipole.

Future work should focus on increasing the realism of the control model by accounting for environmental disturbances and actuator limitations. Refining the tuning strategies for both translational and torsional control—through optimisation or adaptive algorithms—could improve performance under dynamic and uncertain conditions. Additionally, experimental validation, particularly of torsional and coupled vibrations, would strengthen the robustness and practical applicability of the proposed method.

In summary, the contactless control strategy effectively maintains the desired fixed position while attenuating the motion of the magnetic pendulum caused by external pivot disturbances and non-trivial initial conditions. The results highlight the potential of this approach for further development, paving the way toward more intricate multi-degree--of-freedom control algorithms suitable for offshore applications.

Author details

Panagiota Atzampou[1*], Peter C. Meijers[2], Apostolos Tsouvalas[1,2] and Andrei V. Metrikine[1,2]

1 Department of Hydraulic Engineering, Delft University of Technology, Delft, The Netherlands

2 Department of Engineering Structures, Delft University of Technology, Delft, The Netherlands

*Address all correspondence to: p.atzampou@tudelft.nl

References

[1] DOB-Academy: Fox Project Video - Offshore Campaign. Available from: https://www.youtube.com/watch?v=BGcLoREJI-c

[2] Ren Z, Jiang Z, Gao Z, Skjetne R. Active tugger line force control for single blade installation. Wind Energy. 2018; **21**(12):1344-1358. DOI: 10.1002/WE.2258

[3] Jiang Z. Installation of offshore wind turbines: A technical review. Renewable and Sustainable Energy Reviews. 2021; **139**(4):110576. DOI: 10.1016/J.RSER.2020.110576

[4] Neupert J, Mahl T, Haessig B, Sawodny O, Schneider K. A heave compensation approach for offshore cranes. In: Proceedings of the American Control Conference. Netherlands: Elsevier; 2008. pp. 538-543. DOI: 10.1109/ACC.2008.4586547

[5] O'Connor W, Habibi H. Gantry crane control of a double-pendulum, distributed-mass load, using mechanical wave concepts. Mechanical Sciences. 2013;**4**:251-261. DOI: 10.5194/MS-4-251-2013

[6] Atzampou P, Meijers PC, Tsouvalas A, Metrikine AV. Contactless control of suspended loads for offshore installations: Proof of concept using magnetic interaction. Journal of Sound and Vibration. 2024;**575**:118246

[7] Atzampou P, Meijers PC, Tsouvalas A, Metrikine AV. Non-contact electromagnetic control of torsional vibrations of a rigid cylinder. Nonlinear Dynamics. 2025;**113**(3):2001-2016

[8] Bassan M, De Marchi F, Marconi L, Pucacco G, Stanga R, Visco M. Torsion pendulum revisited. Physics Letters A. 2013;**377**(25–27):1555-1562

[9] Blackburn JA, Zhou-Jing Y, Vik S, Smith H, Nerenberg M. Experimental study of chaos in a driven pendulum. Physica D: Nonlinear Phenomena. 1987; **26**(1–3):385-395

[10] Chopra AK. Dynamics of Structures. Hoboken, New Jersey, USA: Wiley, John Wiley & Sons, Ltd.; 2007

Section 3

Mathematical Analysis and Challenges

Chapter 6

In-Depth Analysis of Caputo-Fabrizio Fractional Derivatives: Theory, Applications, and Numerical Scheme

Tarekegn Dinku Gelu, Boka Kumsa and Jyotrimoy Rana

Abstract

This chapter provides a comprehensive exploration of Caputo-Fabrizio fractional derivatives, detailing their theoretical foundations and practical applications. It begins with the properties and definitions of the Caputo-Fabrizio derivative, followed by an analysis of its applications to polynomial, exponential, and trigonometric functions. The discussion encompasses the existence and uniqueness of solutions for fractional differential equations involving this derivative, as well as various numerical schemes designed for their resolution. Additionally, the chapter contrasts integer-order growth models with fractional-order models, highlighting the advantages of fractional derivatives in capturing complex dynamics. A significant application in cancer modeling is also presented, demonstrating the potential of Caputo-Fabrizio derivatives to enhance our understanding of tumor growth dynamics and bridging theoretical concepts with the practical implementations.

Keywords: Caputo-Fabrizio fractional derivative, fractional differential equations, cancer growth model, Laplace transform, fixed point theory

1. Introduction

Fractional calculus is a generalization of classical integration and differentiation to non-integer-order integration and differentiation. It addresses both real and complex orders and their corresponding derivatives. When considering the first and second derivatives, fractional derivatives bridge the gap between these two orders. This approach highlights the significance of higher-order derivatives in understanding complex behaviors a cross various fields. The concept of fractional calculus opens new avenues for analysis and modeling, potentially leading to more accurate predictions and insights. Fractional derivatives can be visualized as curves that lie between the first and second derivatives, providing a continuous transition between these two rates of change. This fractional-order derivative generalizes the classical order derivative using the Cauchy iterative integral [1].

The fractional calculus can be regarded as an ancient yet novel field that has captured the attention of many mathematicians due to its applications in complex and dynamical systems. It is considered an old topic, originating from the early speculations of G.W. Leibniz in 1695 [2]. The discipline of fractional calculus, which involves the assignment of fractional values and their derivatives, was initially introduced by Leibniz as a mathematical curiosity. However, it has since evolved into a significant area of study. The incorporation of fractional calculations into mainstream mathematics was facilitated by the contributions of renowned mathematicians such as Riemann, Liouville, Grunwald, Euler, Lagrange, Caputo, and others, who laid the foundation for contemporary theory [3].

Fractional calculus allows for the computation of integrals and derivatives of any positive real order. It can be considered a branch of mathematical analysis that deals with integro-differential operators and equations, where the integrals are of convolution type and the kernels are of power-law, exponential, or Mittag-Leffler types [4]. The Gamma function and the Mittag-Leffler function [5] are the essential components of fractional calculus, providing the mathematical foundation necessary to define and calculate fractional integrals and derivatives. These concepts are increasingly prevalent across various fields, including physics, engineering, and finance. By utilizing these functions, researchers can effectively model complex systems that traditional integer-order calculus cannot adequately represent, leading to more precise and informative analyses of dynamic behaviors.

Recent studies underscore the growing significance of fractional differential equations in contemporary scientific advancements, particularly in fields such as machine learning, fluid dynamics, and signal processing. The mathematical modeling of real-world problems often leads to the formulation of fractional differential equations. These equations have gained considerable importance due to their diverse applications across multiple disciplines. The behavior of many physical systems can be accurately described within the context of applied sciences and engineering described using fractional-order system theory. Fractional derivatives serve as an excellent tool for the analysis of dynamic systems description of the memory and hereditary properties of various materials and processes, along with the advantages of real objects of the advantages of fractional-order systems that they provide greater degrees of freedom in modeling and incorporate memory effects of the model.

Recently, significant effort has been devoted over the past few years to developing robust and stable numerical and analytical methods for solving fractional differential equations of physical interest. The Adomian decomposition method, homotopy perturbation method, homotopy analysis method, differential transform method, and variational iteration method are relatively new approaches that provide analytical approximate solutions to linear equations.

Fractional calculus is significant for analyzing variations in results between two integers. One of the primary advantages of fractional differential equations (FDE) in mathematical modeling is their non-local property and the memory effect of fractional operators [3]. While integer-order differential operators are local, fractional operators are non-local in that the future state of a system depends not only on its current state but also on all preceding states, due to the convolution-type kernel present in these operators [6, 7]. The term is defined as the dependence of the current output on the historical variations of the input over a finite (or infinite) time interval. For the first time, Boltzmann provided a framework for processes with memory. He proposed a model for isotropic viscoelastic media, where the stress at time t depends not only on the strains at the present moment but also on the historical behavior of the process for $\tau < t$. In 1932,Volterra further developed Boltzmann's theory. Later, the utilization of memory in these processes became a central focus of research, contributing to the development of numerous stochastic models. Granger highlighted the significance of long-range time dependence for the first time in the field of economics.

1.1 Background of fractional calculus

The origins of fractional calculus can be traced back to 1695, with a letter from L'Hôpital. In Leibniz's paper on classical calculus, he discussed linear functions $f(x)$ and the notation $\frac{d^n f(x)}{dx^n}$. The question posed by L'Hôpital was, what if $n = \frac{1}{2}$? The response from Leibniz was, paradox from which 1 day a useful consequence will be drawn. Many mathematicians contributed to the development of fractional calculus [2]. The remarkable journey of fractional calculus began with influential mathematicians such as Euler (1730), P.S. Laplace (1812), J.B.J. Fourier (1822), N.H. Abel (1823), J. Liouville (1832), B. Riemann (1847), K. Grunwald (1867), A.V. Letnikov (1868), Hadamard (1892), M. Riesz (1948), and Caputo (1967) [8–11]. More recently, contributors such as Hataf, M. Fabrizio, Atangana, and Beleanu have made significant advancements in this field. Some of the commonly used fractional-order derivatives today include the Riemann-Liouville (RL), Caputo (C), Caputo-Fabrizio (CF), and Atangana-Beleanu in the Caputo sense (ABC) fractional derivatives.

Fractional derivative categorized based on their kernel can be classified into two type: singular kernel and non-singular kernel. The fractional derivatives like Riemann-Liouville and Caputo employ singular kernel and the Fractional derivatives like Caputo-Fabrizio and Antanga Beleanu utilize non-singular Kernel. In general, they employ power kernel [12], exponential kernel [13], and Mittag-Leffler kernel [14]. Here, power kernel related to Riemann-Liouville and Caputo fractional derivative, the exponential kernel $\exp\left(\frac{-\alpha(t-s)}{1-\alpha}\right)$ is associated with Caputo-Fabrizio fractional derivative and the Mittag-Leffler $E_\alpha\left(\frac{-\alpha(t-s)^\alpha}{1-\alpha}\right)$ corresponds to Atangana-Belanu fractional derivative. In **Figure 1**, the fractional derivative of x^2 approaches to integer-order derivative as α tends to 1 and it coincides with integer order as $\alpha = 1$. These fractional derivatives are based on Cauchy iterative integration. In classical (integer-order) integration, the definite integral in interval $(0, x), x \in \mathbb{R}$ is given by [15]

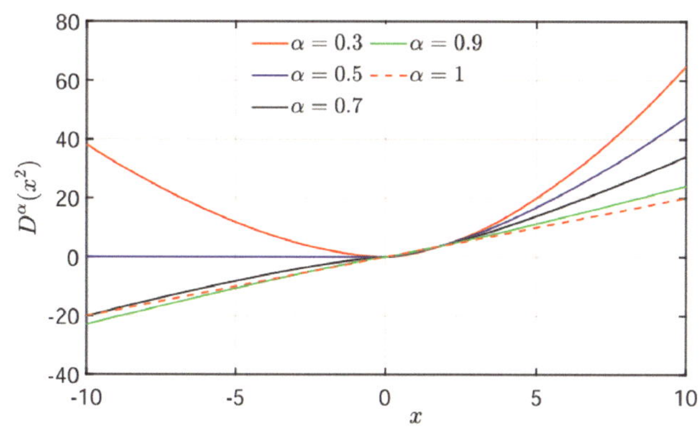

Figure 1.
Caputo-Fabrizio fractional derivative of function $f(x) = x^2$ for different order fractional derivatives.

$$(Jf(x)) = \int_0^x f(s)ds.$$

where J denotes integration. The integral of $(Jf(x))$ is given as

$$(J^2)f(x) = \int_0^x (Jf(t))dt = \int_0^x \left(\int_0^x f(s)ds\right)dt.$$

Iteratively, it follows

$$(J^n f(x)) = \frac{1}{(n-1)!}\int_0^x (x-t)^{n-1} f(t)dt.$$

Now, extending this integer-order integral into fractional order, the generalization of factorial function which is gamma function is applied. Thus, the fractional-order derivative generalized from Cauchy iterative integral is given by

$$J^\alpha f(x) = \frac{1}{\Gamma(\alpha)}\int_0^x (x-t)^{\alpha-1} f(t)dt.$$

Using this fractional integral different mathematicians are tried to extend different fractional-order derivative depending up on their nature of kernel, their initial value problem, and boundary value problem [16–22].

2. Fractional differential operators

2.1 Riemann-Liouville fractional derivative and integral

Let α be real non-negative integer. Let f be piece-wise continuous on $I = (0, \infty)$ and integrable on any finite sub-interval of $I = (0, \infty)$. Then for $t > 0$, the Riemann-Liouville fractional integral is given by [23]

$$I^\alpha f(x) = \frac{1}{\Gamma(\alpha)}\int_c^x (x-t)^{\alpha-1} f(t)dt.$$

For any $c \in I$. This is direct consequences of Cauchy iterative integral.

3. Riemann-Liouville fractional integral of some common functions

1. **Constant function** $f(x) = C$:

$${}_aI_x^\alpha(C) = \frac{1}{\Gamma(\alpha)}\int_a^x (x-t)^{\alpha-1} C dt = \frac{C}{\Gamma(\alpha)}\int_c^x (x-t)^{\alpha-1}.$$

Letting $u = (x-t) \Rightarrow dt = -du$. Then, it follows

$${}_aI_x^\alpha(C) = \frac{C}{\Gamma(\alpha)}\int_{x-a}^0 (u)^{\alpha-1}(-du) = \frac{C}{\alpha\Gamma(\alpha)}(x-a)^\alpha.$$

In particular, if $a = 0$ and $\alpha = 1$ the integration becomes integer order, that is ${}_0I_x^1(C) = Cx = \int_0^x C = 0$. But what if $\alpha = 0$ and $\alpha = 1/2$? For $\alpha = 1/2$, we have

$$ {}_0I_x^{1/2}(C) = \frac{C}{1/2\Gamma(1/2)} x^{1/2} = \frac{C}{1/2\sqrt{\pi}}\sqrt{x} = 2C\sqrt{\frac{x}{\pi}}. $$

2. **Polynomial function** $f(x) = x^n$:

$$ {}_aI_x^\alpha(x^n) = \frac{1}{\Gamma(\alpha)}\int_a^x (x-t)^{\alpha-1} t^n dt = \frac{1}{\Gamma(\alpha)}\int_a^x x^{\alpha-1}\left(1-\frac{t}{x}\right)^{\alpha-1} t^n dt = \frac{x^{\alpha-1}}{\Gamma(\alpha)}\int_a^x \left(1-\frac{t}{x}\right)^{\alpha-1} t^n dt. $$

Now, letting $u = \frac{t}{x}$ and $a = 0$ we have,

$$ {}_0I_x^\alpha(x^n) = \frac{x^{\alpha-1}x^{n+1}}{\Gamma(\alpha)}\int_0^1 (1-u)^{\alpha-1} u du = \frac{x^{\alpha+n}}{\Gamma(\alpha)}B(n+1,\alpha) = \frac{\Gamma(n+1)}{\Gamma(\alpha+n+1)}x^{n+\alpha}. $$

Using $\alpha = n = \frac{1}{2}$, it follows

$$ {}_aI_x^{1/2}\left(x^{1/2}\right) = \frac{\Gamma(1/2+1)}{\Gamma(1/2+1/2+1)}x^{1/2+1/2} = \frac{1/2\Gamma(1/2)}{\Gamma(2)}x = 1/2\sqrt{\pi x}. $$

3. **Exponential functions** $f(x) = e^{kx}$:

$$ {}_aI_x^\alpha\left(e^{kx}\right) = \frac{1}{\Gamma(\alpha)}\int_a^x (x-t)^{\alpha-1} e^{kt} dt. $$

Letting $x - t = u \Rightarrow dt = -du$, it yields

$$ {}_aI_x^\alpha e^{kx} = -\frac{1}{\Gamma(\alpha)}\int_{x-a}^0 u^{\alpha-1} e^{k(x-u)} du = \frac{1}{\Gamma(\alpha)} e^{kx}\int_0^{x-a} e^{-ku} u^{\alpha-1} du. $$

Again letting, $ku = y \Rightarrow du = \frac{1}{k}dy$,

$$ {}_aI_x^\alpha e^{kx} = \frac{1}{\Gamma(\alpha)} e^{kx}\int_0^{k(x-a)} e^{-y}\left(\frac{y}{k}\right)^{\alpha-1}\frac{1}{k}dy = \frac{e^{kx}}{k^\alpha\Gamma(\alpha)}\int_0^\infty e^{-y}y^{\alpha-1}dy = \frac{e^{kx}}{k^\alpha}. $$

Therefore, ${}_aI_x^\alpha\left(e^{kx}\right) = \frac{e^{kx}}{k^\alpha}$.

4. **Trigonometric functions:**

A. **Riemann and Liouville fractional of sine function** $f(x) = \sin(x)$:

$$
\begin{aligned}
{}_0I_x^\alpha(\sin x) &= \frac{1}{\Gamma(\alpha)}\int_0^x (x-t)^{\alpha-1}\sin t dt,\\
&= \frac{1}{\Gamma(\alpha)}\int_0^x (x-t)^{\alpha-1}\sum_{k=0}^{\infty}\frac{(-1)^k t^{2k+1}}{(2k+1)!}dt,\\
&= \frac{1}{\Gamma(\alpha)}\sum_{k=0}^{\infty}\frac{(-1)^k}{(2k+1)!}\int_0^x (x-t)^{\alpha-1}t^{2k+1}dt,\\
&= \sum_{k=0}^{\infty}\frac{(-1)^k}{(2k+1)!}{}_0I_x^\alpha\left(x^{2k+1}\right),\\
&= \sum_{k=0}^{\infty}\frac{(-1)^k}{(2k+1)!}\frac{\Gamma(2k+1+1)}{\Gamma(2k+2+\alpha+1)}x^{2k+1+\alpha} = \frac{(-1)^k}{\Gamma(2k+2+\alpha)}x^{2k+\alpha+1}.
\end{aligned}
$$

B.Riemann and Liouville fractional of cosine function $f(x) = \cos(x)$**:**

$$
\begin{aligned}
{}_0I_x^\alpha(\cos x) &= \frac{1}{\Gamma(\alpha)}\int_0^x (x-t)^{\alpha-1}\cos t dt,\\
&= \frac{1}{\Gamma(\alpha)}\int_0^x (x-t)^{\alpha-1}\sum_{k=0}^{\infty}\frac{(-1)^k t^{2k}}{(2k)!}dt,\\
&= \frac{1}{\Gamma(\alpha)}\sum_{k=0}^{\infty}\frac{(-1)^k}{(2k)!}\int_0^x (x-t)^{\alpha-1}t^{2k}dt,\\
&= \sum_{k=0}^{\infty}\frac{(-1)^k}{(2k)!}{}_0I_x^\alpha\left(x^{2k}\right) = \sum_{k=0}^{\infty}\frac{(-1)^k}{(2k)!}\frac{\Gamma(2k+1)}{\Gamma(2k+\alpha+1)}x^{2k+\alpha},\\
&= \frac{(-1)^k}{\Gamma(2k+1+\alpha)}x^{2k+\alpha}.
\end{aligned}
$$

3.1 Riemann-Liouville fractional derivative

Riemann-Liouville fractional integral is given by

$$ {}_aI_x^\alpha = \frac{1}{\Gamma(\alpha)}\int_a^x (x-t)^{\alpha-1}f(t)dt. $$

And Riemann-Liouville fractional derivative on this integral is given by

$$
\begin{aligned}
{}_aD_x^\alpha f(x) &= \frac{d^n}{dx^n}{}_aI_x^{n-\alpha}f(x),\\
&= \frac{d^n}{dx^n}\left(\frac{1}{\Gamma(n-\alpha)}\int_a^x (x-t)^{n-\alpha-1}f(t)dt\right).
\end{aligned}
$$

Definition 2.1 Riemann-Liouville fractional integral of order $\alpha > 0$ of function $f(t)$ is defined by

$$^{RL}I^{\alpha}f(t) = \int_a^t \frac{(t-s)^{\alpha-1}}{\Gamma(\alpha)} f(s)ds$$

and the fractional derivative of order $\alpha \in [n-1, n]$ of $f(t)$ defined by

$$D^{\alpha}f(t) = I^{n-\alpha}D^{n}f(t), D = \frac{d}{dt}.$$

3.2 Caputo fractional derivative and integral

Let n be the smallest integer that exceeds $\alpha > 0$, then Caputo fractional derivatives of order α is defined as [24]

$$^{C}D^{\alpha}f(x) = \frac{1}{n-\alpha}\int_0^x (x-s)^{n-\alpha-1} f^{n}(s)ds.$$

and

$$^{C}D_x^{\alpha}f(x) = {}_aI_x^{n-\alpha}\left\{\frac{d^n}{dx^n}f(x)\right\}.$$

where

$$^{C}I^{\alpha}f(x) = \frac{1}{\alpha}\int_0^x f(s)(x-s)^{\alpha-1}ds.$$

Example: Evaluate Riemann-Liouville and Capto fractional derivative for $f(x) = C$.

Solution:

I.Let first apply Riemann-Liouville fractional derivative

$$\begin{aligned}
{}_aD_x^{\alpha}C &= \frac{d^n}{dx^n}\,{}_aI_x^{n-\alpha}C = \frac{d^n}{dx^n}\left(\frac{1}{\Gamma(n-\alpha)}\int_a^x (x-t)^{n-\alpha-1}Cdt\right), \\
&= \frac{d^n}{dx^n}\frac{C}{\Gamma(n-\alpha)}\cdot\frac{(x-t)^{n-\alpha}}{n-\alpha}\Big]_a^x, \\
&= \frac{C}{(n-\alpha)\Gamma(n-\alpha)}\left\{\frac{d^n}{dx^n}(x-a)^{n-\alpha}\right\}.
\end{aligned}$$

Using the property $\Gamma(n+1) = n\Gamma(n)$ and $\frac{d^n}{dx^n}x^n = \frac{n!}{(n-m)!}x^{n-m}$ we have

$$\begin{aligned}
\frac{C}{(n-\alpha)\Gamma(n-\alpha)}\frac{d^n}{dx^n}(x-a)^{n-\alpha} &= \frac{(n-\alpha)!}{(n-\alpha-n)}(x-a)^{n-\alpha-n}, \\
&= \frac{C}{\alpha(1+n-\alpha)}\frac{\Gamma(1+n-\alpha)}{(-\alpha)!}(x-a)^{-\alpha} = \frac{C(x-a)^{-\alpha}}{\Gamma(1-\alpha)}, \\
&= \frac{C(x-a)^{-\alpha}}{\Gamma(1-\alpha)}.
\end{aligned}$$

II. Next, let's compute Caputo derivative

$$
\begin{aligned}
{}_aD_x^\alpha(C) = {}_aI_x^{n-\alpha}\left\{\frac{d^n}{dx^n}(C)\right\}, &= \frac{1}{\Gamma(n-\alpha)}\frac{a}{x}(x-t)^{n-\alpha-1}\left[\frac{d^n}{dx^n}(C)\right]dt,\\
&= \frac{1}{\Gamma(n-\alpha)}\frac{a}{x}(x-t)^{n-\alpha-1}(0)dt = 0.
\end{aligned}
$$

Therefore, the Caputo fractional derivative of constant function is zero as classical calculus ${}_aD_x^\alpha C = 0$.

Example: Compute the Caputo fractional derivative of $f(x) = x^\beta$

$$
\begin{aligned}
{}_aD_x^\alpha\left(x^\beta\right) = {}_aI_x^\alpha\left[\frac{d^n}{dx^n}x^\beta\right] &= \frac{1}{n-\alpha}\int_a^x (x-t)^{n-\alpha-1}\frac{d^n}{dt^n}t^\beta dt,\\
&= \frac{1}{n-\alpha}\int_a^x (x-t)^{n-\alpha-1}\frac{\beta!}{(\beta-n)!}t^{\beta-n}dt,\\
&= \frac{1}{n-\alpha}\frac{\beta!}{(\beta-n)!}\int_a^x (x-t)^{n-\alpha-1}t^{\beta-n}dt,\\
&= \frac{1}{n-\alpha}\frac{\beta!}{(\beta-n)!}\int_a^x x^{n-\alpha-1}\left(1-\frac{t}{x}\right)^{n-\alpha-1}t^{\beta-n}dt,\\
y &= \frac{t}{x} \Rightarrow t = xy, dt = xdy,\\
&= \frac{1}{\Gamma(n-\alpha)}\frac{\beta!}{(\beta-n)!}x^{n-\alpha-1}\int_{\frac{a}{x}}^1 (1-y)^{n-\alpha-1}x^{\beta-n}xdy,\\
&= \frac{1}{\Gamma(n-\alpha)}\frac{\beta!}{(\beta-n)!}x^{\beta-\alpha}\int_0^1 (1-y)^{n-\alpha-1}y^{\beta-n}dy,\\
&= \frac{\Gamma(\beta+1)x^{\beta-\alpha}}{\Gamma(n-\alpha)\Gamma(1+\beta-n)}B(n-\alpha, \beta-n+1),\\
&= \frac{\Gamma(\beta+1)x^{\beta-\alpha}}{\Gamma(n-\alpha)\Gamma(1+\beta-n)}\frac{\Gamma(n-\alpha)\Gamma(\beta-n+1)}{\Gamma(\beta-\alpha+1)}x^{\beta-\alpha},\\
&= \frac{\Gamma(1+\beta)}{\Gamma(\beta-\alpha+1)}x^{\beta-\alpha}.
\end{aligned}
$$

Therefore, the Caputo fractional derivative of ${}_aD_x^\alpha\left(x^\beta\right) = \frac{\Gamma(1+\beta)}{\Gamma(\beta-\alpha+1)}x^{\beta-\alpha}$.

Example: Compute Caputo derivative of ${}_aD_x^\alpha e^{\lambda t}$?

$$
\begin{aligned}
{}_aD_x^\alpha e^{\lambda t} = {}_aI_x^{n-\alpha}\left[\frac{d^n}{dx^n}e^{\lambda t}\right], &= {}_aI_x^{n-\alpha}\left[\lambda^n e^{\lambda t}\right],\\
&= \frac{1}{\Gamma(n-\alpha)}\int_a^x (x-t)^{n-\alpha-1}\lambda^n e^{\lambda t}dt,\\
&= \frac{\lambda^n}{\Gamma(n-\alpha)}\int_a^x (x-t)^{n-\alpha-1}\sum_{k=0}^{\infty}\frac{(\lambda t)^k}{k!}dt,\\
&= \frac{\lambda^n}{\Gamma(n-\alpha)}\sum_{k=0}^{\infty}\frac{(\lambda)^k}{k!}\int_a^x (x-t)^{n-\alpha-1}t^k dt,\\
&= \frac{\lambda^n}{\Gamma(n-\alpha)}\sum_{k=0}^{\infty}\frac{(\lambda)^k}{k!}\int_a^x \left(1-\frac{t}{x}\right)^{n-\alpha-1}t^k dt,
\end{aligned}
$$

DOI: http://dx.doi.org/10.5772/intechopen.1013399

Putting $\frac{t}{x} = y \Rightarrow t = xy$ and $dt = xdy$, it follows

$$= \frac{\lambda^n}{\Gamma(n-\alpha)} \sum_{k=0}^{\infty} \frac{(\lambda)^k}{k!} \int_{\frac{a}{x}}^{1} (x)^{n-\alpha-1} (1-y)^{n-\alpha-1} (xy)^k x dy,$$

$$= \frac{\lambda^n}{\Gamma(n-\alpha)} \sum_{k=0}^{\infty} \frac{(\lambda)^k}{k!} (x)^{n-\alpha-1+k+1} \int_{\frac{a}{x}}^{1} (x)^{n-\alpha-1} (1-y)^{n-\alpha-1} (y)^k dy,$$

$$= \frac{\lambda^n}{\Gamma(n-\alpha)} \sum_{k=0}^{\infty} \frac{(\lambda)^k}{k!} (x)^{n-\alpha+k} \beta(n-\alpha, k+1),$$

$$= \frac{\lambda^n}{\Gamma(n-\alpha)} \sum_{k=0}^{\infty} \frac{(\lambda)^k}{k!} (x)^{n-\alpha+k} \frac{\Gamma(n-\alpha)\Gamma(k+1)}{\Gamma(n-\alpha+k+1)},$$

$$= \lambda^n x^{n-\alpha} \sum_{k=0}^{\infty} \frac{(\lambda x)^k}{\Gamma(k+n-\alpha+1)},$$

$$= \lambda^n x^{n-\alpha} E_{1,n-\alpha+1}(\lambda x).$$

The main difference between Riemann-Liouville and Caputo fractional derivative is that

$$^{RL}D^{\alpha} \mathrm{f}(\mathrm{x}) = \frac{d^n}{dx^n} [I^{\alpha}].$$

$$^{C}D^{\alpha} f(x) = I^{n-\alpha} \left[\frac{d^n}{dx^n} f(x) \right].$$

4. Caputo-Fabrizio fractional derivative and integral

Definition 3.1 Let $f \in H'(a,b)$ and $b > a$ and $\alpha \in [0,1]$. Then Caputo fractional derivative is given by [25]

$$^{C}_{b}D^{\alpha}_{a} f(t) = \frac{1}{\Gamma(1-\alpha)} \int_a^t f(s)(t-s)^{-\alpha} ds$$

In both Riemann-Liouville and Caputo fractional derivative uses the kernes $(t-s)^{-\alpha}$, which is singular at $t = s$. Moreover in both Riemann-Liouville and Capauto case the derivative of constant function is not zero. Considering these cases and some initial problems that appear in natural phenomena, Caputo and Fabrizio [13] extended the model using non-singular kernel. In this extension, the used $\exp\left(\frac{-\alpha}{1-\alpha}(t-s)\right)$ instead of $(t-s)^{-\alpha}$ and $\frac{m(\alpha)}{1-\alpha}$ instead of $\frac{1}{\Gamma(1-\alpha)}$. where $m(\alpha)$ is normalization function which satisfies $m(0) = m(1) = 1$. $m(\alpha)$ may takes of form $m(\alpha) = 1 - \frac{1}{\Gamma(\alpha)}$. Using the non-singular kernel(exponential kernel) and the normalization function, the Caputo Fabrizio fractional derivative is given by

$$^{CF}D^{\alpha} f(t) = \frac{m(\alpha)}{1-\alpha} \int_0^t f' s \exp\left(\frac{-\alpha}{1-\alpha}(t-s)\right), \tag{1}$$

where $f \in H'(a,b)$ and $\alpha \in (0,1)$.

Theorem 3.1 prove that

$$\lim_{\alpha\to 1^-}\left({}_t^{CF}D_a^{\alpha}f(x)\right) = f(t) - f(a).$$

Proof: Consider kernel as

$$K_\alpha = \frac{1}{1-\alpha}\exp\left(\frac{-\alpha}{1-\alpha}(x-t)\right).$$

1. **Case 1:** Consider when $x = t$. In this case

$$K_\alpha = \frac{1}{1-\alpha},$$

and

$$\lim_{\alpha\to 1^-} K_\alpha = \lim_{\alpha\to 1^-}\left(\frac{1}{1-\alpha}\right) = \infty$$

2. **Case 2:** When $x > t$, we have $x - t > 0$. that is

$$K_\alpha = \frac{1}{1-\alpha}\exp\left(\frac{-\alpha}{1-\alpha}(x-t)\right),$$

Now taking limit it follows,

$$\lim_{\alpha\to 1^-}\frac{1}{1-\alpha}\exp\left(\frac{-\alpha}{1-\alpha}(x-t)\right) = \frac{\lim_{\alpha\to 1^-}\frac{1}{1-\alpha}}{\lim_{\alpha\to 1^-}\exp\left(\frac{\alpha}{1-\alpha}(x-t)\right)} = \frac{\infty}{\infty},$$

using L'Hospital's rule, we have

$$\lim_{\alpha\to 1^-} K_\alpha = \lim_{\alpha\to 1^-}\left(\frac{1}{(x-t)\frac{1}{(1-\alpha)^3}\exp\left(\frac{\alpha}{1-\alpha}(x-t)\right)}\right) = 0$$

From cases 1 and 2 we have

$$\lim_{\alpha\to 1^-} K_\alpha = \left(\begin{array}{l}\infty, x = t\\ 0, x > t\end{array}\right. = \sigma(x-t),$$

where σ is Kronecker delta function and we know that

$$\int_{-\infty}^{\infty}\sigma(x-t) = 1.$$

From definition of Caputo-Fabrizio,

$$\lim_{\alpha\to 1^-}{}^{CF}D^{\alpha}f(x) = \int_a^x \lim_{\alpha\to 1^-} K_\alpha f'(t) = \int_a^x \sigma(x-t)f'(f) = f'(t).$$

To compare the Caputo-Fabrizio with zero derivative, we use integration by parts as

$$\frac{1}{1-\alpha}\int_0^x f'(t)\exp\left(\frac{-\alpha}{1-\alpha}(x-t)\right)dt = \frac{1}{1-\alpha}\int_0^x f(t)\exp\left(\frac{-\alpha}{1-\alpha}(x-t)\right)\Big|_a^t - \int_a^x f(t)\left(\frac{\alpha}{1-\alpha}\right)\exp\left(\frac{-\alpha(x-t)}{1-\alpha}\right),$$

$$= \frac{f(x)}{1-\alpha} - \frac{f(a)}{1-\alpha} - \frac{\alpha}{1-\alpha}\int_a^x \exp\left(\frac{-\alpha}{1-\alpha}\right)f(t)dt.$$

Applying limit both sides, we have

$$\lim_{\alpha\to 0} {}^{CF}D^{\alpha}f(x) = f(x) - f(a).$$

Caputo-Fabrizio fractional derivative coincides with classical derivative if $f(a) = 0$ with $a = 0$.

5. Examples of Caputo Fabrizio fractional derivatives

Example 3.1 Find CF derivative of $f(t) = t$?
Solution

$$\begin{aligned} {}^{CF}D^{\alpha}t &= \frac{m(\alpha)}{1-\alpha}\int_0^t (t)'\exp\left(\frac{-\alpha}{1-\alpha}(t-s)\right)ds, \\ &= \frac{m(\alpha)}{1-\alpha}\int_0^t \exp\left(\frac{-\alpha}{1-\alpha}(t-s)\right)ds, \\ &= \frac{m(\alpha)}{\alpha}\left(1-e^{\frac{-\alpha t}{1-\alpha}}\right). \end{aligned}$$

Example 3.2 Find the CF derivative of $f(x) = \sin\omega x$?
Solution

$$\begin{aligned} {}^{CF}D^{\alpha}(\sin\omega x) &= \frac{m(\alpha)}{1-\alpha}\int_0^t \omega\cos\omega s\exp\left(\frac{-\alpha}{1-\alpha}(t-s)\right), \\ &= \frac{m(\alpha)}{1-\alpha}\omega\int_0^t \cos\omega s\exp\left(\frac{-\alpha t}{1-\alpha}\cdot\exp\left(\frac{-\alpha s}{1-\alpha}\right)ds,\right. \\ &= \frac{m(\alpha)}{1-\alpha}\omega\exp\left(\frac{-\alpha t}{1-\alpha}\int_0^t \cos\omega s\exp\left(\frac{-\alpha s}{1-\alpha}\right)ds.\right. \end{aligned}$$

Now applying integration by parts,

$$\int_0^t \cos\omega s ds\exp(as) = \frac{1}{a}\cos\omega t\exp at - \frac{1}{a} + \frac{\omega}{a}\int_0^t \cos\omega s\exp(as)ds, \quad a = \frac{\alpha}{1-\alpha},$$

$$\int_0^t \cos\omega s\exp(as)ds = \frac{1}{a}e^{at}\sin\omega t - \frac{\omega}{a}\int_0^t \cos\omega s ds.$$

$$\int_0^t \cos\omega s ds = \frac{\left(\frac{1}{a}e^{at}\left(\cos\left(\omega t + \frac{\omega}{a}\sin(\omega t)\right)\right) - \frac{1}{a}\right)}{1+\frac{\omega^2}{a^2}}.$$

Thus,

$$ {}^{CF}D^{\alpha}(\sin\omega x) = \frac{m(\alpha)}{1-\alpha}\omega\exp\left(\frac{-\alpha t}{1-\alpha}\right)\left[\frac{\left(\frac{1}{\frac{\alpha}{1-\alpha}}e^{\frac{\alpha}{1-\alpha}t}\left(\cos\left(\omega t+\frac{\omega}{\frac{\alpha}{1-\alpha}}\sin(\omega t)\right)\right)-\frac{1}{\frac{\alpha}{1-\alpha}}\right)}{1+\frac{\omega^2}{\left(\frac{\alpha}{1-\alpha}\right)^2}}\right]. $$

Example 3.3 Find CF derivative of $f(t)=e^{\omega t}$?
Solution

$$ \begin{aligned} {}^{CF}D^{\alpha}e^{\omega t} &= \frac{m(\alpha)}{1-\alpha}\int_0^t (e^{\omega t})'\exp\left(\frac{-\alpha}{1-\alpha}(t-s)\right)ds, \\ &= \frac{m(\alpha)}{1-\alpha}\int_0^t \omega e^{\omega s}e^{-at}e^{at}ds \\ &= \frac{m(\alpha)}{1-\alpha}\omega e^{-at}\int_0^t e^{(\omega+a)s}ds, \\ &= \frac{m(\alpha)}{1-\alpha}\omega e^{-at}\left[\frac{e^{(\omega+a)t}-1}{\omega+a}\right], \\ &= \frac{m(\alpha)}{1-\alpha}\omega e^{-\left(\frac{\alpha}{1-\alpha}\right)t}\left[\frac{e^{\left(\omega+\left(\frac{\alpha}{1-\alpha}\right)\right)t}-1}{\omega+\left(\frac{\alpha}{1-\alpha}\right)}\right]. \end{aligned} $$

Example 3.4 Find the CF derivative of $f(t)=t^{\beta}$.
Solution: By definition of Cafuto Fabrizio derivative, it follows

$$ \begin{aligned} {}^{CF}D^{\alpha}t^{\beta} &= \frac{m(\alpha)}{1-\alpha}\int_0^t (s^{\beta})'\exp\left(\frac{-\alpha}{1-\alpha}(t-s)\right)ds, \\ &= \frac{m(\alpha)}{1-\alpha}\int_0^t \beta s^{\beta-1}\exp\left(\frac{-\alpha t}{1-\alpha}\right)\exp\left(\frac{\alpha s}{1-\alpha}\right)ds, \\ &= \beta\frac{m(\alpha)}{1-\alpha}\exp\left(\frac{-\alpha t}{1-\alpha}\right)\int_0^t s^{\beta-1}\exp\left(\frac{\alpha s}{1-\alpha}\right)ds. \end{aligned} $$

Now, let's evaluate integral

$$ \int_0^t s^{\beta-1}\exp\left(\frac{\alpha s}{1-\alpha}\right)ds. $$

If $\beta=1$, the integral reduces to $\int_0^t \exp\left(\frac{\alpha s}{1-\alpha}\right)ds$, which can be evaluated directly:

$$ \frac{1-\alpha}{\alpha}\left[\exp\left(\frac{\alpha t}{1-\alpha}\right)-1\right]. $$

If $\alpha=0$, the integral becomes $\int_0^t s^{\beta-1}ds=\frac{t^{\beta}}{\beta}$. Using Taylor series expansion, we have

$$ \exp\left(\frac{\alpha s}{1-\alpha}\right)=\sum_{n=0}^{\infty}\frac{(\alpha s)^n}{(1-\alpha)^n n!}, $$

and integrating term by term we have

$$\int_0^t s^{\beta-1}\sum_{n=0}^{\infty}\frac{(\alpha s)^n}{(1-\alpha)^n n!}ds=\sum_{n=0}^{\infty}\frac{\alpha^n}{(1-\alpha)^n n!}\int_0^t s^{\beta+n-1}ds=\sum_{n=0}^{\infty}\frac{\alpha^n t^{\beta+n}}{(1-\alpha)^n n!(\beta+n)}.$$

$$\exp\left(\frac{\alpha s}{1-\alpha}\right)=\sum_{n=0}^{\infty}\frac{(\alpha s)^n}{(1-\alpha)^n n!},$$

and integrate term by term:

$$\int_0^t s^{\beta-1}\sum_{n=0}^{\infty}\frac{(\alpha s)^n}{(1-\alpha)^n n!}ds=\sum_{n=0}^{\infty}\frac{\alpha^n}{(1-\alpha)^n n!}\int_0^t s^{\beta+n-1}ds=\sum_{n=0}^{\infty}\frac{\alpha^n t^{\beta+n}}{(1-\alpha)^n n!(\beta+n)}.$$

This series converges for all $t\geq 0$ and $\alpha\in(0,1)$. Thus

$${}^{CF}D^{\alpha}t^{\beta}=\beta\frac{m(\alpha)}{1-\alpha}\exp\left(\frac{-\alpha t}{1-\alpha}\right)\left[\sum_{n=0}^{\infty}\frac{\alpha^n t^{\beta+n}}{(1-\alpha)^n n!(\beta+n)}\right].$$

6. Difference between Riemann-Liouville, and Caputo and Caputo-Fabrizio fractional derivatives

The primary distinction among these three types of fractional integrals lies in their treatment of initial conditions. The Riemann-Liouville fractional integral is defined using the Riemann-Liouville derivative, which necessitates that the function be differentiable of integer order at the initial point. In contrast, the Caputo fractional integral employs the Caputo derivative, a modification of the Riemann-Liouville derivative that accommodates nonzero initial conditions. The Caputo-Fabrizio fractional integral represents a further refinement of the Caputo derivative, incorporating a weighted average of the function and its derivatives over a finite interval. Fractional integrals are generalizations of classical integer-order integrals, and their definitions depend on the type of fractional derivative employed. The three most common types—Riemann-Liouville, Caputo, and Caputo-Fabrizio—primarily differ in their treatment of initial conditions, the structure of their kernels, and their applicability for modeling physical process (**Table 1**).

Feature	Riemann-Liouville	Caputo	Caputo-Fabrizio
Kernel type	Singular power-law	Singular power-law	Non-singular exponential
Initial conditions	Fractional-order	Integer-order	Integer-order
Physical interpretability	Limited	Good	Very good
Regularity requirement	Differentiable	Differentiable	Continuous
Application focus	Theoretical, mathematical	Physical, engineering	Real-world, complex systems
Memory sescription	Long memory (power-law)	Long memory (power-law)	Fading memory (exponential)

Table 1.
Comparison of fractional derivative approaches [26–28].

i. *Riemann-Liouville fractional integral*

This fractional integral is based on the Riemann-Liouville fractional derivative. It employs a singular kernel, typically of the form $(t-s)^{-\alpha}$, where $\alpha > 0$. This integral requires initial conditions that involve fractional derivatives that are not always physically interpretable and can often be difficult to measure. The function must be at least locally integrable and often needs to be differentiable of integer order at the initial point. Commonly used in pure mathematics and theoretical models of anomalous diffusion and memory effects in viscoelastic materials, the non-local and singular kernel may pose challenges in numerical simulations and physical interpretation.

ii. *Caputo fractional integral*

This fractional integral is derived from the Caputo fractional derivative, which modifies the Riemann-Liouville approach to enhance its applicability. It also employs a singular power-law kernel, $(t-s)^{\alpha-1}$, but applies it in a manner that shifts the differentiation after integration. This method requires only integer-order derivatives at the initial point, making it more suitable for physical problems where these conditions are known. The function should be differentiable of integer order; however, the fractional derivative remains well-defined for less smooth functions. It is widely used in engineering, biology, and physics, particularly in systems with memory, such as tumor growth models, drug kinetics, and heat conduction with memory. It offers better compatibility with real-world initial conditions compared to the Riemann-Liouville approach.

iii. *Caputo-Fabrizio fractional integral*

It is based on the Caputo-Fabrizio derivative, which replaces the singular kernel with a non-singular alternative. Its kernel is a non-singular exponential kernel, typically of the form $e^{\frac{-\alpha(t-s)}{1-\alpha}}$. It is similar to Caputo in that it employs integer-order initial conditions, which are more practical to implement. This approach is less restrictive, as it requires only that functions be continuous. It is particularly well-suited for modeling real-world phenomena characterized by fading memory, such as biological systems, economic models, electrical circuits, and epidemiological studies. It is easier to implement numerically due to the smooth kernel and has greater physical relevance for processes with exponentially decaying memory.

6.1 Laplace transform of Caputo-Fabrizio fractional derivative

The Laplace transform of function f in the interval $(0, \infty)$ is given by [29]

$$\mathcal{L}(f(t)) = \int_0^\infty e^{-st} f(t)dt$$

and

$$\mathcal{L}(f'(t)) = s\mathcal{L}(f(t)) - f(0).$$

In case of Caputo-Fabrizio fractional derivative,

$$\mathcal{L}\left({}^{CF}D^{\alpha}\right)f(x) = \mathcal{L}\left(\frac{1}{1-\alpha}\int_a^x f'(t)\exp\left(\frac{-\alpha}{1-\alpha}(x-t)\right)dt\right), \tag{2}$$

$$= \frac{1}{1-\alpha}\mathcal{L}\left(\int_a^x f'(t)\exp\left(\frac{-\alpha}{1-\alpha}(x-t)\right)dt\right). \tag{3}$$

Recall that the convolution two of functions $f(t)$ and $g(t)$ is given by

$$f^*g = \int_a^t f(x-t)g(x)dx$$

and its Laplace transform is given by

$$\mathcal{L}(f^*g) = \mathcal{L}(f(t)) \times \mathcal{L}(g(t)).$$

Now, from the definition of Convolution,

$$\exp\left(\frac{-\alpha}{1-\alpha}x\right) * f'(x) = \int_a^x f'(t)\exp\left(\frac{-\alpha}{1-\alpha}(x-t)\right)dt.$$

Thus, the Laplace transform of

$$\mathcal{L}\left(\int_a^x f'(t)\exp\left(\frac{-\alpha}{1-\alpha}(x-t)\right)dt\right) = \mathcal{L}\left(\exp\left(\frac{-\alpha}{1-\alpha}(x)\right) \times \mathcal{L}(f'(x)),\right. \tag{4}$$

$$= \frac{1}{s+\frac{\alpha}{1-\alpha}} \times (s\mathcal{L}(f(t)) - f(0)). \tag{5}$$

Therefore,

$$\mathcal{L}\left({}^{CF}D^{\alpha}\right)f(x) = \frac{1}{1-\alpha}\left(\frac{1}{s+\frac{\alpha}{1-\alpha}} \times (s\mathcal{L}(f(t)) - f(0))\right), \tag{6}$$

$$= \frac{sF(s) - f(0)}{s+\alpha(1-s)}. \tag{7}$$

where $F(s) = \mathcal{L}(f(t))$.

6.2 Caputo-Fabrizio fractional integral

Let $0<\alpha<1$. Consider the following non-homogeneous fractional differential equation:

$${}^{cF}D^{\alpha}f(t) = u(t), t \geq 0. \tag{8}$$

Applying Laplace transform both sides of Eq. (8), we obtain

$$\frac{s\mathcal{L}(f(t)) - f(0)}{s+\alpha(1-s)} = \mathcal{L}(u(t)).$$

Rearranging the equation, we have

$$\mathcal{L}(f(t)) = \frac{f(0)}{s} + (1-\alpha)\mathcal{L}(u(t)) + \alpha\frac{\mathcal{L}(u(t))}{s}. \tag{9}$$

Now, applying inverse Laplace transform both sides of Eq. (9), we get [17]

$$f(t) = f(0) + (1-\alpha)u(t) + \alpha\int_0^t u(s)ds, \tag{10}$$

$$= (1-\alpha)u(t) + \alpha\int_0^t u(s)ds + c, \tag{11}$$

is solution of problem (8), where c is integral constant. Thus, the associate Caputo-Fabrizio fractional integral is given by

$$^{CF}I^{\alpha}f(t) = (1-\alpha)u(t) + \alpha\int_0^t u(s)ds. \tag{12}$$

we have done the integration with process called normalization function $m(\alpha) = 1$. We can choose $m(\alpha)$, so that

$$(1-\alpha)m(\alpha) + \alpha m(\alpha) = 1.$$

Theorem 5.1 [30] Let $u(t) \in \mathbb{R}^n$ and $\lambda, \eta \in \mathbb{R}$, then the solution of Cauchy problem

$$\begin{cases} ^{CF}D^{\alpha}u(t) = \lambda u(t), \\ u(t_0) = \eta, \end{cases}$$

is given by

$$u(t) = \frac{\eta}{1+\lambda(\alpha-1)}\exp\left(\frac{\lambda\alpha t}{1+\lambda(\alpha-1)}\right).$$

Proof. Applying Laplace transform to both sides of Cauchy Problem, we have

$$\frac{su(s)-u(0)}{s+\alpha(1-s)} = \lambda u(s) \Rightarrow u(s) = \frac{\eta}{1+\lambda(\alpha-s)}\left(\frac{eta}{s-\frac{\lambda\alpha}{1+\lambda(\alpha-1)}}\right).$$

Now, applying inverse Laplace transform, it results

$$u(t) = \frac{\eta}{1+\lambda(\alpha-s)}\exp\left(\frac{\lambda\alpha t}{1+\lambda(\alpha-1)}\right).$$

If $\frac{\lambda\alpha t}{1+\lambda(\alpha-1)} < 0$, then all the solutions converges.

6.3 Non-homogeneous fractional differential equation involving Caputo-Fabrizio fractional-order derivative with known function $g(t)$

Consider [30] the problem with CF fractional order derivative with $u \in H^1(a,b)$ and $g(t) \in C^1(a,b)$

$$^{CF}D^{\alpha}u(t) = u(t) + g(t) \tag{13}$$

Then, the solution of this non-Homogeneous Cauchy problem is given by

$$\begin{aligned}u(t) &= u(t) + u(t)\\ &= \frac{u(0)M(\alpha)}{M(\alpha)+\alpha-1}\exp\left(\frac{\alpha t}{M(\alpha)+\alpha-1}\right),\\ &+\left(\frac{1-\alpha}{M(\alpha)+\alpha-1}\right)g(t)+\left(\frac{1-\alpha}{M(\alpha)+\alpha-1}\right)\left(\frac{\alpha}{1-\alpha}+,\right.\\ &\left.\frac{1}{M(\alpha)+\alpha-1}\right)\int_0^t g(\zeta)\exp\left(\frac{t-\zeta}{M(\alpha)+\alpha-1}\right)d\zeta.\end{aligned}$$

Proof. Applying definition of Caputo-Fabrication fractional derivative on left hand side, it follows

$$\frac{M(\alpha)}{1-\alpha}\int_0^t u'(\zeta)\exp\left(\frac{-\alpha(t-\zeta)}{1-\alpha}\right)dt = u(t)+g(t).$$

Applying Laplace transform both sides, we have

$$\frac{M(\alpha)}{1-\alpha}\left[\frac{su(s)-u(0)}{s+(1-\alpha)s} = u(s)+g(s)\right].$$

Rearranging, we have

$$\begin{aligned}u(s) &= \frac{M(\alpha)}{1-\alpha}\cdot\frac{u(0)}{s+\frac{\alpha}{1-\alpha}}\left[\frac{s+\frac{\alpha}{1-\alpha}}{s\left[\frac{M(\alpha)}{1-\alpha}-1\right]-\frac{\alpha}{1-\alpha}}\right]+\frac{g(s)\left(s+\frac{\alpha}{1-\alpha}\right)}{s\left[\frac{M(\alpha)}{1-\alpha}-1\right]-\frac{\alpha}{1-\alpha}},\\ &= \frac{M(\alpha)}{1-\alpha}\left[\frac{u(0)}{s\left[\frac{M(\alpha)}{1-\alpha}-1\right]-\frac{\alpha}{1-\alpha}}\right]+\frac{g(s)\left(s+\frac{\alpha}{1-\alpha}\right)}{s\left[\frac{M(\alpha)}{1-\alpha}-1\right]-\frac{\alpha}{1-\alpha}}.\end{aligned}$$

Splitting $u(s) = u_1(s) + u_2(s)$ into two parts where,

$$u_1(s) = \frac{M(\alpha)}{1-\alpha}\left[\frac{u(0)}{s\left[\frac{M(\alpha)}{1-\alpha}-1\right]-\frac{\alpha}{1-\alpha}}\right],\quad u_2(s) = \frac{g(s)\left(s+\frac{\alpha}{1-\alpha}\right)}{s\left[\frac{M(\alpha)}{1-\alpha}-1\right]-\frac{\alpha}{1-\alpha}}.$$

Now applying inverse Laplace transform we have,

$$\begin{aligned}\mathbb{L}^{-1}(u_1(s)) &= \mathbb{L}^{-1}\left(\frac{M(\alpha)}{1-\alpha}\left[\frac{u(0)}{s\left[\frac{M(\alpha)}{1-\alpha}-1\right]-\frac{\alpha}{1-\alpha}}\right]\right),\\ &= \frac{u(0)M(\alpha)}{M(\alpha)+\alpha-1}\mathbb{L}^{-1}\left\{\frac{1}{s-\frac{\alpha}{M(\alpha)+\alpha-1}}\right\}.\end{aligned}$$

Thus,

$$u_1(t) = \frac{u(0)M(\alpha)}{M(\alpha)+\alpha-1}\exp\left(\frac{\alpha t}{M(\alpha)+\alpha-1}\right).$$

Secondly, applying inverse Laplace transform on $u_2(t)$, we have

$$\mathbb{L}^{-1}(u_2(s)) = \mathbb{L}^{-1}\left(\frac{g(s)\left(s+\frac{\alpha}{1-\alpha}\right)}{s\left[\frac{M(\alpha)}{1-\alpha}-1\right]-\frac{\alpha}{1-\alpha}}\right),$$

$$= \frac{1-\alpha}{M(\alpha)+\alpha-1}\mathbb{L}^{-1}\left(g(s)+\frac{g(s)\left(\frac{\alpha}{1-\alpha}+\frac{1}{M(\alpha+\alpha-1)}\right)}{1-\frac{1}{M(\alpha)+\alpha-1}}\right).$$

$$u_2(t) = \left(\frac{1-\alpha}{M(\alpha)+\alpha-1}\right)g(t)+$$
$$\left(\frac{1-\alpha}{M(\alpha)+\alpha-1}\right)\left(\frac{\alpha}{1-\alpha}+\frac{1}{M(\alpha)+\alpha-1}\right)\int_0^t g(\zeta)\exp\left(\frac{t-\zeta}{M(\alpha)+\alpha-1}\right)d\zeta.$$

Thus,

$$u(t) = u(t)+u(t)$$
$$= \frac{u(0)M(\alpha)}{M(\alpha)+\alpha-1}\exp\left(\frac{\alpha t}{M(\alpha)+\alpha-1}\right),$$
$$+\left(\frac{1-\alpha}{M(\alpha)+\alpha-1}\right)g(t)+\left(\frac{1-\alpha}{M(\alpha)+\alpha-1}\right)\left(\frac{\alpha}{1-\alpha}+,\right.$$
$$\left.\frac{1}{M(\alpha)+\alpha-1}\right)\int_0^t g(\zeta)\exp\left(\frac{t-\zeta}{M(\alpha)+\alpha-1}\right)d\zeta.$$

6.4 Existence and uniqueness of Caputo-Fabrizio fractional derivatives

The existence and uniqueness of CF has been introduced using Picard's successive iteration approach with help of Banach fixed point theorem.

Theorem 6.1 [31] Let A be an region on which a function is defined and $A = \{(t,y) : |t-t_0| \leq a, |y-y_0| \leq b\}$. Moreover, suppose function $f : A \to \mathbb{R}$, is continuous inside A and satisfies Lipschitz condition with respect to second variable. Then the initial value problem

$$^{CF}D^{\alpha}y = f(t,y(t)), \ \ y(0) = y_0 \tag{14}$$

has unique solution. Where $^{CF}D^{\alpha}$ is Caputo-Fabrizio fractional derivative of order $\alpha \in (0,1)$.

Proof. Applying fundamental theorem of fractional calculus, it follows

$$y(t) = y(0) + {}^{CF}I^{\alpha}(f(s,y(s))) \tag{15}$$

where $^{CF}I^{\alpha}$ is Caputo Fabrizio fractional integral of order α. Now, applying Picards successive iteration on Eq. (15), it follows

$$y_n(t) = y(0) + {}^{CF}I^{\alpha}(f(s,y_n(s))) \tag{16}$$

with $y_0(t) \equiv y_0$. But from the definition of Caputo-Fabrizio fractional integral with

$$^{CF}I^{\alpha}f(t,y(t)) = \frac{2(1-\alpha)}{(2-\alpha)m(\alpha)}f(t,y(t)) + \frac{2\alpha}{(2-\alpha)m(\alpha)}\int_{t_0}^{t} f(s,y(s))ds.$$

Thus, the Picards iteration of CF takes of form

$$y_n(t) = y(0) + \frac{2(1-\alpha)}{(2-\alpha)m(\alpha)}f(t,y_{n-1}(t)) + \frac{2\alpha}{(2-\alpha)m(\alpha)}\int_{t_0}^{t} f(s,y_{n-1}(s))ds. \tag{17}$$

Using the successive approximation method, we show that the sequence defined by

$$\begin{aligned}
y_0(t) &= y_0,\\
y_1(t) &= y_0 + \frac{2(1-\alpha)}{(2-\alpha)m(\alpha)}f(t,y_0(t)) + \frac{2\alpha}{(2-\alpha)m(\alpha)}\int_{t_0}^{t} f(s,y_0(s))ds,\\
y_2(t) &= y_0 + \frac{2(1-\alpha)}{(2-\alpha)m(\alpha)}f(t,y_1(t)) + \frac{2\alpha}{(2-\alpha)m(\alpha)}\int_{t_0}^{t} f(s,y_1(s))ds,\\
&\vdots,\\
y_n(t) &= y(0) + \frac{2(1-\alpha)}{(2-\alpha)m(\alpha)}f(t,y_{n-1}(t)) + \frac{2\alpha}{(2-\alpha)m(\alpha)}\int_{t_0}^{t} f(s,y_{n-1}(s))ds.
\end{aligned}$$

To ensure that Picards iteration is well defined, it is sufficient to ensure each $y_n(t)$ lies inside region A. That is,

$$\|y_n(t) - y_0\| \le b, \quad |t - t_0| \le h. \tag{18}$$

We proof Eq. (18) using mathematical induction. Suppose $M = \max_{(t,y)\in A} f(t,y(t))$.

Proof: For $n = 1$, $y_1(t) = y_0 + \frac{2(1-\alpha)}{(2-\alpha)m(\alpha)}f(t,y_0(t)) + \frac{2\alpha}{(2-\alpha)m(\alpha)}\int_{t_0}^{t} f(s,y_0(s))ds$.

$$\begin{aligned}
\|y_1 - y_0\| &= \left\| \frac{2(1-\alpha)}{(2-\alpha)m(\alpha)}f(t,y_0(t)) + \frac{2\alpha}{(2-\alpha)m(\alpha)}\int_{t_0}^{t} f(s,y_0(s))ds \right\|,\\
&\le \frac{2(1-\alpha)}{(2-\alpha)m(\alpha)}\|f(t,y_0(t))\| + \frac{2\alpha}{(2-\alpha)m(\alpha)}\int_{t_0}^{t} \|f(s,y_0(s))ds\|,\\
&\le \frac{2(1-\alpha)}{(2-\alpha)m(\alpha)}M + \frac{2\alpha}{(2-\alpha)m(\alpha)}M\int_{t_0}^{t} ds,\\
&= M\left(\frac{2(1-\alpha)}{(2-\alpha)m(\alpha)} + \frac{2\alpha}{(2-\alpha)m(\alpha)}(t-t_0)\right),\\
&\le Mh \le b, t \to t_0.
\end{aligned}$$

Assume the assertion is true for $n = i$, That is,

$$\|y_i(t) - y_0\| \le Mh \le b$$

Now, we want to show its true for $n = i + 1$.

$$\|y_{i+1} - y_0\| = \left\| \frac{2(1-\alpha)}{(2-\alpha)m(\alpha)} f(t, y_i(t)) + \frac{2\alpha}{(2-\alpha)m(\alpha)} \int_{t_0}^{t} f(s, y_i(s)) ds \right\|, \tag{19}$$

$$\leq \frac{2(1-\alpha)}{(2-\alpha)m(\alpha)} \|f(t, y_i(t))\| + \frac{2\alpha}{(2-\alpha)m(\alpha)} \int_{t_0}^{t} \|f(s, y_i(s))\| ds, \tag{20}$$

$$\leq \frac{2(1-\alpha)}{(2-\alpha)m(\alpha)} M + \frac{2\alpha}{(2-\alpha)m(\alpha)} M \int_{t_0}^{t} ds, \tag{21}$$

$$\leq Mh \leq b, t \to t_0. \tag{22}$$

Thus, Picards iteration of CF is well-defined in region A.

Next, we will show sequence of Picards iteration converges uniformly in the interval $|t - t_0| \leq h$. This is accomplished writing the sequences as

$$y_n(t) = y_0(t) + (y_1(t) - y_0) + (y_2(t) - y_1(t)) + \dots + (y_n(t) - y_{n-1}(t)).$$

Clearly, the sequence $y_n(t)$ converges, if and only if the series

$$y_0(t) + (y_1(t) - y_0) + (y_2(t) - y_1(t)) + \dots + (y_n(t) - y_{n-1}(t))$$

converges. To proof this it suffices that

$$\sum_{n=1}^{\infty} \|y_{n+1} - y_n\| < \infty.$$

For $t \in (t_0 - h, t_0 + h)$, denote $\|D_n\| = \|y_{n+1} - y_n\|$.

$$\begin{aligned}
\|D_n\| &= \|y_{n+1} - y_n\|, \\
&= \left\| \frac{2(1-\alpha)}{(2-\alpha)m(\alpha)} (f(t, y_n(t) - f(t, y_{n-1}(t)))) + \frac{2\alpha}{(2-\alpha)m(\alpha)} \int_{t_0}^{t} (f(s, y_s(s) - f(s, y_{s-1}(s)))) ds \right\|, \\
&= \frac{2(1-\alpha)}{(2-\alpha)m(\alpha)} L \|y_n(t) - y_{n-1}(t))\| + \frac{2\alpha}{(2-\alpha)m(\alpha)} L \int_{t_0}^{t} \|y_n(s) - y_{n-1}(s)))\| ds, \\
&\leq \frac{2(1-\alpha)}{(2-\alpha)m(\alpha)} L \|D_{n-1}\| + \frac{2\alpha}{(2-\alpha)m(\alpha)} L \|D_{n-1}\| (t - t_0).
\end{aligned}$$

If there exist t^*, such that $\left(\frac{2(1-\alpha)}{(2-\alpha)m(\alpha)} + \frac{2\alpha}{(2-\alpha)m(\alpha)} (t^* - t_0) \right) L < 1$, then there exist unique solution for Eq. (14).

Proof. Observe that,

$$\begin{aligned}
\|y_1 - y_0\| &= \left\| k(\alpha) f(t, y_0) + w(\alpha) \int_{t_0}^{t} f(s, y_0) ds \right\|, \\
&\leq k(\alpha) M + w(\alpha) M (t - t_0), \\
&= M(k(\alpha) + w(\alpha) h), \quad h = t - t_0.
\end{aligned}$$

and

$$\begin{aligned}
\|y_2 - y_1\| &= \left\| k(\alpha)\big(f(t,y_1) - f(t,y_0)\big) + w(\alpha)\int_{t_0}^{t} \big(f(s,y_1) - f(t,y_0)\big)ds \right\|,\\
&\leq k(\alpha)\|f(t,y_1) - f(t,y_0)\| + w(\alpha)\int_{t_0}^{t} \|f(s,y_1) - f(t,y_0)\|ds,\\
&\leq \big(k(\alpha)L + w(\alpha)L(t-t_0)\big)\big(y_1 - y_0\big),\\
&= L\big(k(\alpha) + (\alpha)h\big)\big(y_1 - y_0\big), \quad h = t - t_0,\\
&= L\big(k(\alpha) + (\alpha)h\big)\big(M(k(\alpha) + w(\alpha)h)\big),\\
&= LM\big(k(\alpha) + w(\alpha)h\big)^2.
\end{aligned}$$

Inductively,

$$\|y_n - y_{n-1}\| \leq M\big(k(\alpha) + w(\alpha)h\big)^n L^{n-1}.$$

Thus, if the assumptions holds true then, the series converges uniformly. Thus, $y_n(t) \to y(t)$. Therefore, there exists solution.

To prove the uniqueness of solution, assume $y_1(t)$ and $y_2(t)$ are the two distinct solutions of the initial value problem (14). From the definition, we have

$$\begin{aligned}
\|y_1 - y_2\| &\leq k(\alpha)\|y_1 - y_2\| + w(\alpha)\int_{t_0}^{t} \|y_1 - y_2\|ds,\\
&\leq k(\alpha)\|y_1 - y_2\| + w(\alpha)\|y_1 - y_2\|h,\\
&= \big(k(\alpha) + w(\alpha)h\big)\|y_1 - y_2\|,
\end{aligned}$$

which implies

$$\|y_1 - y_2\|\big(1 - k(\alpha) - w(\alpha)h\big) \leq 0.$$

Therefore,

$$y_1 = y_2.$$

Thus, $y(t)$ is unique solution of the given initial value problem (14).

7. Numerical scheme for Caputo-Fabrizio fractional derivative

7.1 L_1 method for CF

Let us recall Caputo Fabrizio fractional derivative as [32]

$$^{CF}D^{\alpha}f(x) = \frac{m(\alpha)}{1-\alpha}\int_0^x f'(s)\exp(\lambda(x-s))ds, \quad \lambda = \frac{-\alpha}{1-\alpha}. \tag{23}$$

Letting $x = x_n$ in the Eq. (23), we have

$$ {}^{CF}D^{\alpha}f(x_n) = \frac{m(\alpha)}{1-\alpha}\int_0^x f'(s)\exp(\lambda(x_n-s))ds. \tag{24} $$

$$ {}^{CF}D^{\alpha}f(x_n) = \frac{m(\alpha)}{1-\alpha}\left[\int_0^{x_1} f'(s)\exp(\lambda(x_n-s))ds + \int_1^{x_2} f'(s)\exp(\lambda(x_n-s))ds + ,\right. \\ \left. \ldots + \int_{x_{n-1}}^{x_n} f'(s)\exp(\lambda(x_n-s))ds\right], \tag{25} $$

$$ = \frac{m(\alpha)}{1-\alpha}\sum_{k=1}^{n}\int_{x_{k-1}}^{x_k} f'(s)\exp[\lambda(x_n-s)]ds. \tag{26} $$

Now, we replace the Eq. (26) by forward difference as follows:

$$ \begin{aligned} {}^{CF}D^{\alpha}f(x_n) &= \frac{m(\alpha)}{1-\alpha}\sum_{k=1}^{n}\int_{x_{k-1}}^{x_k} \exp[\lambda(x_n-s)]\left[\frac{f(x_k)-f(x_{k-1})}{h}\right]ds, \\ &= \frac{m(\alpha)}{h(1-\alpha)}\sum_{k=1}^{n}\int_{x_{k-1}}^{x_k} \exp[\lambda(x_n-s)][f(x_k)-f(x_{k-1})]ds, \\ &= \frac{m(\alpha)}{h(1-\alpha)}\sum_{k=1}^{n}[f(x_k)-f(x_{k-1})]\int_{x_{k-1}}^{x_k} \exp[\lambda(x_n-s)]ds. \end{aligned} $$

Integrating by power rule, we get

$$ \begin{aligned} {}^{CF}D^{\alpha}f(x_n) &= \frac{m(\alpha)}{h(1-\alpha)}\sum_{k=1}^{n}[f(x_k)-f(x_{k-1})]\left[\exp\left(\frac{\lambda(x_n-s)}{-\lambda}\right)\right]_{x_{k-1}}^{x_k}, \\ &= \frac{m(\alpha)}{\lambda h(1-\alpha)}\sum_{k=1}^{n}[f(x_k)-f(x_{k-1})][\exp(\lambda(x_n-x_k)) - \exp(\lambda(x_n-x_{k-1}))]. \end{aligned} $$

Now, using the relation: $x_n = nh \quad x_{k-1} = (k-1)h, \quad x_k = kh$, the equation becomes:

$$ \begin{aligned} {}^{CF}D^{\alpha}f(x_n) &= \frac{m(\alpha)}{-\lambda h(1-\alpha)}\sum_{k=1}^{n}[f(kh)-f((k-1)h)][\exp(\lambda(nh-kh)) - \exp(\lambda(nh-(k-1)h))], \\ &= \frac{m(\alpha)}{\alpha h}\sum_{k=1}^{n}[f(kh)-f((k-1)h)][\exp(\lambda(nh-kh)) - \exp(\lambda(nh-(k-1)h))]. \end{aligned} $$

where $\lambda = \left(\frac{-\alpha}{1-\alpha}\right)$ and $m(\alpha) = 1$.

7.2 Three-step Adams-Bashforth numerical scheme for Caputo-Fabrizio fractional derivative

Consider the fractional differential equation with Caput-Fabrizio fractional derivative [33]

$$ {}^{CF}D^{\alpha}u(t) = f(t, u(t)). \tag{27} $$

Using the definition of the Caputo Fabrizio fractional derivative, it follows

$$\frac{M(\alpha)}{1-\alpha}\int_0^t u'(\zeta)\exp\left(\frac{-\alpha(t-\zeta)}{1-\alpha}\right)dt = f(t,u(t)).$$

Applying fundamental theorem of fractional calculus, it follows

$$u(t)-u(0) = \frac{1-\alpha}{M(\alpha)}f(t,u(t)) + \frac{\alpha}{M(\alpha)}\int_0^t f(\zeta,u(\zeta))d\zeta,$$

using discretization for each time, we obtain

$$u(t_{n+1})-u(0) = \frac{1-\alpha}{M(\alpha)}f(t_n,u(t_n)) + \frac{\alpha}{M(\alpha)}\int_0^t f(\zeta,u(\zeta))d\zeta, \tag{28}$$

and

$$u(t_n)-u(0) = \frac{1-\alpha}{M(\alpha)}f(t_{n-1},u(t_{n-1})) + \frac{\alpha}{M(\alpha)}\int_0^t f(\zeta,u(\zeta))d\zeta. \tag{29}$$

Subtracting Eq. (28) from (29) it yields

$$u(t_{n+1})-u(t_n) = \frac{1-\alpha}{M(\alpha)}\{f(t_n,u_n) - f(t_{n-1},u(t_{n-1}))\} + \frac{\alpha}{M(\alpha)}\int_{t_n}^{t_{n+1}} f(\zeta,u(\zeta))d\zeta. \tag{30}$$

But

$$\begin{aligned}\int_{t_n}^{t_{n+1}} f(\zeta,u(\zeta))d\zeta &= \int_{t_n}^{t_{n+1}}\left\{\frac{f(t_n,u_n)}{h}(t-t_n) - \frac{f(t_{n-1},u_{n-1})}{h}(t-t_{n-1}) + \frac{f(t_{n-2},u_{n-2})}{h}(t-t_n)dt\right\},\\ &= \frac{23h}{12}f(t_n,u_n) - \frac{16h}{12}f(t_{n-1},u_{n-1}) + \frac{5h}{12}f(t_{n-2},u_{n-2}),\end{aligned}$$

which implies

$$\begin{aligned}u(t_{n+1})-u(t_n) &= \left(\frac{1-\alpha}{M(\alpha)}+\frac{23\alpha h}{12M(\alpha)}\right)f(t_n,u_n) - \left(\frac{1-\alpha}{M(\alpha)}+\frac{16\alpha h}{12M(\alpha)}\right)f(t_{n-1},u_{n-1}),\\ &+\frac{5\alpha h}{12M(\alpha)}f(t_{n-2},u_{n-2}).\end{aligned}$$

Hence,

$$\begin{aligned}u(t_{n+1}) &= u(t_n) + \left(\frac{1-\alpha}{M(\alpha)}+\frac{23\alpha h}{12M(\alpha)}\right)f(t_n,u_n) - \left(\frac{1-\alpha}{M(\alpha)}+\frac{16\alpha h}{12M(\alpha)}\right)f(t_{n-1},u_{n-1})\\ &+\frac{5\alpha h}{12M(\alpha)}f(t_{n-2},u_{n-2})\end{aligned}$$

Clearly, with $\alpha = 1$, we get classical Adams-Bashforth scheme, that is,

$$u_{n+1} = u_n + \frac{h}{12}[23f(t_n,u_n) - 16f(t_{n-1},u_{n-1}) + 5f(t_{n-2},u_{n-2})].$$

7.3 Predictor corrector scheme for Caputo Fabrizio fractional derivative

Consider the fractional differential equation with Caput-Fabrizio fractional derivative [34]

$$^{CF}D^{\alpha}u(t) = f(t, u(t)). \tag{31}$$

Using the definition of the Caputo Fabrizio fractional derivative, it follows

$$\frac{M(\alpha)}{1-\alpha}\int_0^t u'(\zeta)\exp\left(\frac{-\alpha(t-\zeta)}{1-\alpha}\right)dt = f(t, u(t)).$$

Evaluating for each time at two points, with $t_m = m\Delta t,\ \ t_{m+1} = (m+1)\Delta t$, which yields

$$u(t_{m+1} - u(t_m)) = \frac{1-\alpha}{M(\alpha)}(f(t_{m+1}, u(t_{m+1}) - f(t_m, u(t_m)))) + \frac{\alpha}{M(\alpha)}\int_{t_m}^{t_{m+1}} f(\zeta, u(\zeta))d\zeta.$$

Function $f(\zeta, u(\zeta))$ can be approximated over interval $[t_m, t_{m+1}]$ by means of Newton polynomial, we have

$$\begin{aligned} u_{m+1} &= u_m + \frac{\alpha\Delta}{M(\alpha)}f(t_{m+1}, u_{m+1}) + \left(\frac{1-\alpha}{M(\alpha)}\right)(f(t_{m+1}, u(t_{m+1})) - f(t_m, u(t_m))), \\ &+ \frac{\alpha}{M(\alpha)}\left(\frac{f(t_{m+1}, u(t_{m+1})) - f(t_m, u(t_m))}{\Delta t}\right)\int_{t_m}^{t_{m+1}}(\zeta - t_m)d\zeta, \\ &+ \frac{\alpha}{M(\alpha)}\left(\frac{f(t_{m+1}, u(t_{m+1})) - 2f(t_m, u(t_m)) + f(t_{m-1}, u(t_{m-1}))}{2(\Delta t)^2}\right)\int_{t_m}^{t_{m+1}}(\zeta - t_m)(\zeta - t_{m+1})d\zeta. \end{aligned}$$

Replacing the integration from respective, and rearranging,

$$\begin{aligned} u_{m+1} &= u_m + \frac{1-\alpha}{M(\alpha)}\left(f\left(t_{m+1}, u_{m+1}^p\right) - f(t_m, u_m),\right. \\ &+ \frac{\alpha}{M(\alpha)}\left(\frac{5}{12}f\left(t_{m+1}, u_{m+1}^p\right)\Delta t\right) - \frac{2}{3}f(t_m, u_m)\Delta t - f(t_{m-1}, u_{m-1})\left.\frac{\Delta t}{12}\right). \end{aligned}$$

With predictor

$$\begin{aligned} u_{m+1}^p &= u_m + \frac{1-\alpha}{M(\alpha)}\left(f(t_m, u_m) - f(t_{m-1}, u_{m-1}),\right. \\ &+ \frac{\alpha}{M(\alpha)}\left(\frac{5}{12}f(t_{m-2}, u_{m-2})\Delta t\right) - \frac{4}{3}f(t_{m-1}, u_{m-1})\Delta t + f(t_m, u_m)\left.\frac{23\Delta t}{12}.\right) \end{aligned}$$

8. Stability of fractional differential equation involving Caputo-Fabrizio fractional derivative

Consider the fractional differential equation involving Caputo-Fabrizio fractional derivative with linear coefficient.

$$^{CF}D^{\alpha} = Au(t),$$

where $u(t) \in \mathbb{R}^n$ is state variable and A is constant matrix. The solution is stable if the matrix $I_n + (\alpha - 1)A$ is invertible.

Theorem 8.1 [35] Let x^* be steady state of ${}^{CF}D^{\alpha} = Au(t)$. If the matrix $I_n + (\alpha - 1)A$ is invertible and the condition

$$\left| arg\left(\lambda\left(\alpha A(I_n + (\alpha-1)A)^{-1}\right)\right)\right| \geq \frac{\alpha\pi}{2}$$

is hold then the equilibrium point is asymptotically stable.

Proof: Applying Laplace transform both sides, we have

$$\begin{aligned}
&\Rightarrow \frac{su(s) - u(s)}{s + \alpha(1-s)} = Au(s),\\
&\Rightarrow su(s) - \eta = sAu(s) + \alpha(1-s)Au(s),\\
&\Rightarrow [s(I_n + (\alpha-1)A) - \alpha A]u(s) = \eta,\\
&\Rightarrow [s[I_n + (\alpha-1)A] - \alpha A]u(s) = \eta,\\
&\Rightarrow u(s) = \frac{\eta(I_n + (\alpha-1)A)^{-1}}{sI_n - \alpha A(I_n + (\alpha-1)A)^{-1}}.
\end{aligned}$$

Applying inverse Laplace transform both sides, then it yields

$$u(t) = \psi \exp\left(\alpha A(I_n + (\alpha-1)A)^{-1}t\right),$$

where $\psi = \eta(I_n + (\alpha-1)A)^{-1})$ is a vector. This solution $u(t)$ converges to zero if

$$\left| arg\left(\lambda\left(\alpha A(I_n + (\alpha-1)A)^{-1}\right)\right)\right| \geq \frac{\alpha\pi}{2}.$$

Theorem 8.2 [36] Consider the system

$$\begin{cases} {}^{CF}D^{\alpha}u(t) = f(t, u(t)) \\ u(0) = u_0. \end{cases} \tag{32}$$

where, $u \in \mathscr{R}^n$ and $f(t, \cdot) : \mathbb{R}^3 \to \mathbb{R}^3$, for $t \in [0, \tau]$, $\tau > 0$ belongs to $C^1(\mathbb{R}^3)$ and $\alpha \in (0, 1)$. Then, the characteristic equation of Jacobian matrix the system (32) at equilibrium point u^* is given by

$$det(s(I_{n\times n} - (1-\alpha)J(u^*)) - \alpha J(u^*)) = 0, \tag{33}$$

and the system (32) is asymptotically stable if eigenvalues of the characteristic Eq. (33) satisfies one of the following conditions:

$$1. \|\lambda_j\| \geq \frac{1}{1-\alpha}, \quad 2. Re\left(\lambda_j\right) > \frac{1}{1-\alpha}, 3. Re\left(\lambda_j\right) < 0, 4. |Im\left(\lambda_j\right)| > \frac{1}{2(1-\alpha)}.$$

where λ_j are the eigenvalues of Jacobian matrix evaluated at given equilibrium point.

Theorem 8.3 [37] Let u^* be equilibrium point of the non-linear system

$${}^{CF}D^{\alpha}u(t) = Au(t),$$

with Caputo-Fabrizio fractional operator. Then, the equilibrium point u^* is asymptotically stable if the eigenvalues of Jacobian matrix $J(u^*)$ satisfy the following

1. $\|\lambda(J(u^*))| \geq \frac{1}{1-\alpha}$, $\lambda(J(u^*)) \neq \frac{1}{1-\alpha}$,

2. $Re\left(\lambda(J(u^*) > \frac{1}{1-\alpha}\right.$,

3. $Re(\lambda(J(u^*) < 0$,

4. $Re\left(\lambda\left(J(u^*) > \frac{1}{2(1-\alpha)}\right.\right.$.

Proof: Here, we use linearization concept near equilibrium point(perturbation) using Taylor series expansion. Now, consider system of fractional differential equations

$$\begin{aligned} {}^{CF}D^{\alpha}u_1(t) &= f_1(x_1, x_2, \dots, x_n), \\ {}^{CF}D^{\alpha}u_2(t) &= f_2(x_1, x_2, \dots, x_n), \\ &\vdots \\ {}^{CF}D^{\alpha}u_n(t) &= f_n(x_1, x_2, \dots, x_n), \end{aligned}$$

with initial conditions

$$u_{10} = u_{10}, \quad u_{20} = u_{20}, \dots, u_{n0} = u_{n0}.$$

To compute steady states, we set ${}^{CF}D^{\alpha}u_i(t) = 0 \Rightarrow u^* = \left(u_1^*, u_2^*, \dots, u_n^*\right)$. To show asymptotic stability set $u_i(t) = u_i^* + \epsilon_i, i = 1,2,3, \dots, n$. Then

$${}^{CF}D^{\alpha}\left(u_i^* + \epsilon_i\right) = f_i\left(\left(u_1^* + \epsilon_1, u_2^* + \epsilon_2, \dots, u_n^* + \epsilon_n\right),\right.$$

which implies

$${}^{CF}D^{\alpha}(\epsilon_i) = f_i\left(\left(u_1^* + \epsilon_1, u_2^* + \epsilon_2, \dots, u_n^* + \epsilon_n\right).\right.$$

Using Taylor series expansion of f near equilibrium point yields,

$$f_i\left(\left(u_1^* + \epsilon_1, u_2^* + \epsilon_2, \dots, u_n^* + \epsilon_n\right) = f_i\left(\left(u_1^*, u_2^*, \dots, u_n^*\right) + \left.\frac{\partial f_i}{\partial u_1}\right|_* \epsilon_1 + \left.\frac{\partial f_i}{\partial u_2}\right|_* \epsilon_2 + \dots + \left.\frac{\partial f_i}{\partial u_n}\right|_* \epsilon_n.\right.\right.$$

Which implies, ${}^{CF}D^{\alpha}(\epsilon_i) = A\epsilon_i$. where

$$\epsilon = \begin{pmatrix} \epsilon_1 \\ \epsilon_1 \\ \vdots \\ \epsilon_n \end{pmatrix} \quad A = \begin{bmatrix} a_{11} & a_{12} & a_{13} & \dots a_{1n} \\ a_{21} & a_{22} & a_{23} & \dots a_{2n} \\ a_{31} & a_{32} & a_{33} & \dots a_{3n} \\ \vdots & \vdots & \vdots & \vdots \\ a_{n1} & a_{n2} & a_{n3} & \dots a_{nn} \end{bmatrix} a_{ij} = \left.\frac{\partial f_i}{\partial x_j}\right|_*, i, j = 1, 2, \dots, n.$$

Thus, we have

$$B^{-1}AB = C,$$

where C is diagonal matrix of A given by

$$C = \begin{bmatrix} \lambda_1 & 0 & 0 & \ldots 0 \\ 0 & \lambda_2 & 0 & \ldots 0 \\ 0 & 0 & \lambda_3 & \ldots 0 \\ \vdots & \vdots & \vdots & \vdots \\ 0 & 0 & 0 & \ldots \lambda_n \end{bmatrix},$$

where λ_j are Eigen values of A and B is eigenvector of matrix A, then

$$AB = BC, A = BCB^{-1},$$

which implies

$$\begin{aligned} {}^{CF}D^{\alpha}\epsilon &= \left(BCB^{-1}\right)\epsilon, \\ {}^{CF}D^{\alpha}\left(B^{-1}\epsilon\right) &= C\left(B^{-1}\right)\epsilon, \\ {}^{CF}D^{\alpha}\eta &= C\eta, \quad \eta = B^{-1}\epsilon. \end{aligned}$$

Applying Laplace transform with the initial conditions, we obtain,

$$det(s(I-(1-\alpha)C-\alpha C)) = 0.$$

Then, the system is asymptotically stable if and only if the real parts of roots of the characteristic equation is negative.

9. Application of Caputo-Fabrizio fractional derivative on cancer model

The application of the Caputo-Fabrizio fractional derivative in cancer modeling provides a sophisticated approach to understanding tumor growth dynamics and treatment effects [38, 39]. This mathematical framework employs fractional calculus to capture memory effects and non-local interactions [40–43], which are crucial in biological systems. For example, fractional differential equations can effectively model tumor growth, drug delivery dynamics, and interactions between tumor cells and the immune system. By utilizing these models, researchers can gain deeper insights into the complex behavior of cancer, leading to improved therapeutic strategies and a better understanding of tumor-immune interactions [44, 45].

In this study, we present a demonstration of tumor-immune interactions using the fractional-order Caputo-Fabrizio derivative, focusing on the dynamics between active and resting immune cells. The model explores how the activation of immune cells in response to tumor presence can be characterized using fractional derivatives, effectively capturing the delayed and history-dependent nature of immune responses. This approach provides a more nuanced understanding of how different states of immune cells interact with tumor cells over time, offering valuable insights into potential therapeutic interventions that can enhance the efficacy of cancer treatments. The interaction of tumor cells-immune cells with the Michaelis-Menten term and the conversion of the active immune cell into the resting immune cell have been considered. Tumor cells ($T(t)$), hunting immune cells ($H(t)$), and resting or helper immune cells ($R(t)$) are governed by the following system of non-linear differential equation.

$$\begin{cases} \dfrac{dT}{dt} = r_1 T\left(1 - \dfrac{T}{k_1}\right) - \alpha_1 TH + c, \\ \dfrac{dH}{dt} = \beta HR - \sigma_1 H, \\ \dfrac{dR}{dt} = r_2 R\left(1 - \dfrac{R}{k_2}\right) - \rho_1 HR - \rho_2 R + \dfrac{\rho TR}{T+\eta} + \epsilon_1 H, \end{cases} \tag{34}$$

with the initial conditions $T(0) = T_0, H(0) = H_0,$ and $R(0) = R_0$. Here, r_1 and k_1 are the growth rate and the carrying capacity of tumor cells, respectively. r_2 and k_2 denote the resting cells' growth rate and carrying capacity, respectively. The parameter α_1 is the rate of killing tumor cells by hunting immunity cells, and β is the conversion rate of helper T cells (resting) to active immune cells (hunting). On the other hand, ρ_1 represents the conversion rate of the resting immune cells to the hunting immune cells. Furthermore, the parameters c and ρ_2 represent the conversion of the normal cell into malignant tumor cells and apoptosis of resting immune cells, respectively.

Non-dimensionalizing the model, we have

$$\begin{cases} x' = a_0 x(1-x) - xy + 1, \\ y' = a_1 yz - a_2 y, \\ z' = a_3 z(1-z) - a_4 yz - a_5 z + a_6 \dfrac{xz}{x+k} + a_7 y, \end{cases} \tag{35}$$

with $x(0) = x_0, y(0) = y_0,$ and $z(0) = z_0$. Motivated by the above model, we propose the following fractional order differential system in the Caputo-Fabrizio sense in the interval $[0, \tau]$ as:

$$\begin{cases} {}^{CF}D^{\alpha} x = a_0^{\alpha} x(1-x) - xy + 1, \\ {}^{CF}D^{\alpha} y = a_1^{\alpha} yz - a_2^{\alpha} y, \\ {}^{CF}D^{\alpha} z = a_3^{\alpha} z(1-z) - a_4^{\alpha} yz - a_5^{\alpha} z + a_6^{\alpha} \dfrac{xz}{x+k^{\alpha}} + a_7^{\alpha} y, \end{cases} \tag{36}$$

with $x(0) = x_0 \geq 0, y(0) = y_0 \geq 0$ and $z(0) = z_0 \geq 0$ and $\alpha \in (0, 1]$. The definition and results related to the Caputo-Fabrizio fractional order derivative are given in the next section. From Eq. (36), the steady-state equation is given by

$$ {}^{CF}D^{\alpha} X(t) = 0 \tag{37}$$

The system has three equilibrium points. These are

$$E_1 = (x_1, 0, 0), E_2 = (x_2, 0, z_2) \text{ and } E_3 = \left(x_3, y_3, \frac{a_2^{\alpha}}{a_1^{\alpha}}\right),$$

where $x_1 = \frac{1}{2}\left(1 + \sqrt{1 + \frac{4}{a_0^{\alpha}}}\right), x_2 = \frac{1}{2}\left(1 + \sqrt{1 + \frac{4}{a_0^{\alpha}}}\right), z_2 = \frac{1}{a_3^{\alpha}}\left(a_3^{\alpha} - a_5^{\alpha} + \frac{a_6^{\alpha} x_2}{x_2 + k^{\alpha}}\right)$.

Moreover, E_1 is exists always and E_2 exists if $\left(a_3^{\alpha} + \frac{a_6^{\alpha} x_2}{x_2 + k^{\alpha}}\right) > a_5^{\alpha}$,

$$y_3 = \frac{b}{a} + \frac{c}{a}(x_3/(x_3 + k^\alpha))$$

with

$$a = \frac{a_4^\alpha a_2^\alpha}{a_1^\alpha}, b = \frac{a_3^\alpha a_2^\alpha}{a_1^\alpha} - \frac{a_3^\alpha a_2^{2\alpha}}{a_1^{2\alpha}} - \frac{a_5^\alpha a_2^\alpha}{a_1^\alpha}, c = \frac{a_6^\alpha a_2^\alpha}{a_1^\alpha}$$

and x_3 is positive real root of

$$Q_1 x_3^3 + Q_2 x_3^2 + Q_3 x_3 + Q_4 = 0, \tag{38}$$

where

$$Q_1 = a_0^\alpha > 0, Q_2 = a_0^\alpha k^\alpha - a_0^\alpha + b/a + c/a, Q_3 = (bk^\alpha/a) - a_0^\alpha k^\alpha - 1, Q_4 = -k^\alpha < 0.$$

From the Descartes rule of sign change [46], we can confirm the existence of a positive real root x_3. From the first equation of (37) we have $y_3 = \frac{a_0^\alpha x_3(1-x_3)+1}{x_3}$, which implies $y_3 > 0$ if $a_0^\alpha x_3(1-x_3) + 1 > 0$. Thus, there exist endemic positive equilibrium point $E_3 = (x_3, y_3, z_3)$.

In order to check the stability of system (36), we use classical Jacobian matrix concept. The Jacobian matrix of (36) at $E_i, i = 1,2,3.$ is given by

$$J(E_i)_= \begin{bmatrix} a_0^\alpha - 2a_0^\alpha x_i - y_i & -x_i & 0 \\ 0 & a_1^\alpha z_i - a_2^\alpha & a_1^\alpha y_i \\ \frac{a_6^\alpha k^\alpha z_i}{(x_i + k^\alpha)^2} & -a_4^\alpha z_i - a_7^\alpha & a_3^\alpha - 2a_3^\alpha z_i - a_4^\alpha y_i - a_5^\alpha + \frac{a_6^\alpha x_i}{x_i + k^\alpha}. \end{bmatrix} \tag{39}$$

Now, we describe the condition for the stability of the system (36) in the following theorems.

Theorem 9.1 The equilibrium point E_1 is locally asymptotically stable if $a_3^\alpha + \frac{a_6^\alpha x_1}{x_1+k^\alpha} < a_5^\alpha$ or $a_3^\alpha + \frac{a_6^\alpha x_1}{x_1+k^\alpha} - a_5^\alpha > \frac{1}{\alpha^2+1}$.

Proof:

$$J(E_1)_= \begin{bmatrix} -a_0^\alpha\sqrt{1+\frac{4}{a_0^\alpha}} & -x_1 & 0 \\ 0 & -a_2^\alpha & 0 \\ 0 & a_7^\alpha & a_3^\alpha - a_5^\alpha + \frac{a_6^\alpha x_1}{x_1 + k^\alpha}. \end{bmatrix}$$

Then, the characteristic equation is given by

$$det(s(I_{3\times3} - (1-\alpha)J(E_1)) - \alpha J(E_1)) = 0$$

Let

$$e_1 = -a_0^\alpha\sqrt{1+\frac{4}{a_0^\alpha}}, e_2 = -x_1, e_2 = -a_2^\alpha, e_4 = a_7^\alpha, e_5 = a_3^\alpha - a_5^\alpha + \frac{a_6^\alpha x_1}{x_1 + k^\alpha},$$

$$\mathcal{E} = s(I_{3\times 3} - (1-\alpha)J(E_1)) - \alpha J(E_1).$$

Then

$$\mathcal{E} = \begin{bmatrix} s(1-(1-\alpha)e_1) - \alpha e_1 & -s(1-\alpha)e_2 - \alpha e_2 & 0 \\ 0 & s(1-(1-\alpha)e_3) - \alpha e_3 & 0 \\ 0 & -s(1-\alpha)e_4 - \alpha e_4 & s(1-\alpha)e_5 - \alpha e_5. \end{bmatrix}$$

The characteristic equation of $\mathcal{E}$ is given by

$$(s(1-(1-\alpha)e_1) - \alpha e_1)(s(1-(1-\alpha)e_3) - \alpha e_3)(s(1-\alpha)e_5 - \alpha e_5) = 0$$

From the equation, we get the following eigenvalues:

$$s_1 = \frac{\alpha e_1}{1-(1-\alpha)e_1}, s_2 = \frac{\alpha e_3}{1-(1-\alpha)e_3}, s_3 = \frac{\alpha e_5}{1-(1-\alpha)e_5}$$

Since e_1 and e_3 have negative real part, it stratifies the third condition of stability. The eigenvalue s_3 satisfies the condition if $e_5 < 0$ or it satisfy $s_3 > \frac{1}{1-\alpha}$. That is $e_5 > \frac{1}{\alpha^2+1}$. Thus, if s_3 satisfies either of the conditions system(4) asymptotically stable. Otherwise unstable.

Theorem 9.2 The equilibrium point E_2 is stable if $a_3^\alpha + \frac{a_6^\alpha x_1}{x_1+k^\alpha} > a_5^\alpha$ or $\left(-a_3^\alpha - \frac{a_6^\alpha x_1}{x_1+k^\alpha} + a_5^\alpha\right) > \frac{1}{\alpha}$ holds

Proof:

$$J(E_2) = \begin{bmatrix} -a_0^\alpha\sqrt{1+\frac{4}{a_0^\alpha}} & -x_2 & 0 \\ 0 & a_1^\alpha z_2 - a_2^\alpha & 0 \\ \frac{a_6^\alpha k^\alpha}{(x_2+K^\alpha)^2} & -a_4^\alpha - a_7^\alpha & -a_3^\alpha - \frac{a_6^\alpha x_1}{x_1+k^\alpha} + a_5^\alpha. \end{bmatrix}$$

Let $b_1 = -a_0^\alpha\sqrt{1+\frac{4}{a_0^\alpha}}, b_2 = a_1^\alpha z_2 - a_2^\alpha, b_3 = \frac{a_6^\alpha k^\alpha}{(x_2+K^\alpha)^2}, b_4 = -a_4^\alpha - a_7^\alpha, b_5 = -a_3^\alpha - \frac{a_6^\alpha x_1}{x_1+k^\alpha} + a_5^\alpha$ Then, the characteristic equation is given by

$$det(s(I_{3\times 3} - (1-\alpha)J(E_2)) - \alpha J(E_2)) = 0$$

which is equivalent to

$$\begin{vmatrix} s(1-(1-\alpha)b_1) - \alpha b_1 & s(1-\alpha)x_2 + \alpha x_2 & 0 \\ 0 & s(1-(1-\alpha)b_2) - \alpha b_2 & 0 \\ -s(1-\alpha)b_3 - \alpha b_3 & -s(1-\alpha)b_4 - \alpha b_4 & s(1-(1-\alpha)b_5) - \alpha b_5 \end{vmatrix} = 0$$

Now, it follows,

$$(s((1-\alpha)b_1 - \alpha b_1))(s(1-(1-\alpha)b_2) - \alpha b_2)(s(1-(1-\alpha)b_5) - \alpha b_5)) = 0$$

solving the equation, we get

$$s_1 = \frac{\alpha b_1}{1-(1-\alpha)b_1}, s_2 = \frac{\alpha b_2}{1-(1-\alpha)b_2}, s_3 = \frac{\alpha b_5}{1-(1-\alpha)b_5}.$$

It clear that the real part of eigenvalue s_1 and s_2 is negative. The system to be stable at E_2 either $b_5 < 0.$ or $b_5 > \frac{1}{1-\alpha}$.

To study the stability of the endemic equilibrium point E_3, we will utilize the results of the discriminant of polynomial order 3 and the Routh-Hurwitz criteria for stability. The Jacobian matrix of system (36) at E_3 is given by

$$J(E_3)_= \begin{bmatrix} \omega_1 & \omega_2 & 0 \\ 0 & \omega_3 & \omega_4 \\ \omega_5 & \omega_6 & \omega_7 \end{bmatrix} \tag{40}$$

where

$$\omega_1 = a_0^\alpha - 2a_0^\alpha x_3 - y_3, \quad \omega_2 = -x_3, \omega_3 = a_1^\alpha z_3 - a_2^\alpha, \omega_4 = a_1^\alpha y_3, \omega_5 = \frac{a_6^\alpha k^\alpha z_3}{(x_3+k^\alpha)^2},$$

$$\omega_6 = -a_4^\alpha z_3 - a_7^\alpha, \omega_7 = a_3^\alpha - 2a_3^\alpha z_3 - a_4^\alpha y_3 - a_5^\alpha + \frac{a_6^\alpha x_3}{x_3+k^\alpha}.$$

Then, the characteristic equation is given by

$$det(s(I_{3\times 3} - (1-\alpha)J(E_3)) - \alpha J(E_3)) = 0$$

which is equivalent to

$$\begin{vmatrix} s(1-(1-\alpha)w_1) - \alpha w_1 & s(1-\alpha)x_2 + \alpha x_2 & 0 \\ 0 & s & s(1-\alpha)w_4 + \alpha w_4 \\ -s(1-\alpha)w_5 - \alpha w_5 & -s(1-\alpha)_w 6 - \alpha w_6 & s(1-(1-\alpha)w_7) - \alpha w_7 \end{vmatrix} = 0$$

Let $1-(1-\alpha)w_1 = d_1, (1-\alpha)w_2 = d_2, (1-\alpha)w_3 = d_3, (1-\alpha)w_5 = d_4, (1-\alpha)w_6 = d_5$ and $1-(1-\alpha)w_7 = d_6$ Then, the determinant computed as

$$(d_1 s - \alpha w_1)\left(d_6 s^2 - \alpha s w_7 - (d_5 s - \alpha w_6)(d_3 s - \alpha w_4)\right)$$
$$+(d_2 s - \alpha w_2)(d_4 s - \alpha w_5)(d_3 s - \alpha w_4) = 0$$

After simplifying we get

$$p_1 s^3 + p_2 s^2 + p_3 s + p_4 = 0, \tag{41}$$

where

$$p_1 = d_1 d_6 + d_2 d_3 d_4, p_2 = -(\alpha d_6 w_1 + d_3 d_5 + \alpha d_2 d_4 w_4 + \alpha d_2 d_3 w_5 + \alpha d_1 w_7 + \alpha d_3 d_d w_2),$$

$$p_3 = \alpha^2 w_1 w_7 + \alpha d_5 w_4 + \alpha^2 d_2 w_4 w_5 + \alpha d_3 d_6 + \alpha^2 w_2 w_4 d_4 + \alpha^2 d_3 w_2 w_5,$$

$$p_4 = \alpha^2 w_6 w_4 - \alpha^3 w_2 w_4 w_5.$$

From Eq. (41), we have negative real root if

$$\frac{p_2}{p_1}, \frac{p_{23}}{p_1}, \frac{p_4}{p_1} > 0, \text{and } \frac{p_2 p_3}{p_1^2} > \frac{p_4}{p_1}.$$

Therefore, the endemic equilibrium point is has negative real root Routh-Huritz criteria. Moreover, if we Let

$$\lambda = s - \frac{a}{3}, a = \frac{p_2}{p_1}, b = \frac{p_3}{p_1}, c = \frac{p_4}{p_1},$$

then the cubic equation reduces and takes of form

$$\lambda^2 + p\lambda + q = 0, \text{where } p = \frac{3b - a^2}{3}, q = \frac{2a^3 - 9ab + 27c}{27}.$$

The discriminant of the cubic polynomial is given by $\Delta = \left(\frac{q}{2}\right)^2 + \left(\frac{p}{3}\right)^3$. Depending on the sign of the discriminant, we have the following roots for Eq. (41):

- If $\Delta = 0$, then all roots are real and given by

$$s_1 = 2\left(\frac{-q}{2}\right)^{\frac{1}{3}} - \frac{a}{3}, s_{2,3} = -\left(\frac{-q}{2}\right)^{\frac{1}{3}} - \frac{a}{3}$$

- If $\Delta > 0$, then we have one real and two complex roots and given by

$$s_1 = \left(-\frac{q}{2} + \sqrt{\Delta}\right)^{\frac{1}{3}} - \left(\frac{q}{2} + \sqrt{\Delta}\right)^{\frac{1}{3}} - \frac{a}{3},$$

$$s_{2,3} = \frac{-1}{2}(u_1 - v_1) \pm (u_1 + v_1)\frac{\sqrt{3}}{2}i - \frac{a}{3},$$

where

$$u_1 = \left(\frac{-q}{2}\right)^{\frac{1}{3}}, \quad v_1 = \left(\frac{q}{2}\right)^{\frac{1}{3}}.$$

- If $\Delta < 0$, then we have the following roots

$$s_1 = 2\sqrt{\left(\frac{-p}{3}\right)^3} \cos\frac{\phi}{3} - \frac{a}{3}, s_2 = 2\sqrt{\left(\frac{-p}{3}\right)^3} \cos\frac{\phi + 2\pi}{3} - \frac{a}{3},$$

$$s_3 = 2\sqrt{\left(\frac{-p}{3}\right)^3} \cos\frac{\phi + 4\pi}{3} - \frac{a}{3}, \text{where, } \cos\phi = \frac{-q}{2\left(\frac{-p}{3}\right)^3}$$

If either of the Routh-Hurwitz criteria or all $s_1, s_2, s_3 > \frac{1}{1-\alpha}$ is satisfied then the system is asymptotically stable at E_3.

10. Numerical results

In this section, we compare the numerical results of the Caputo fractional order derivative, the Caputo-Fabrizio fractional-order derivative, and the Atangana-Belanu fractional-order derivative across various fractional orders. In **Figure 2(a)-(c)**, it is observed that the Caputo-Fabrizio fractional derivative acts as an intermediary between the Caputo fractional derivative and the Atangana-Belanu fractional derivative. In particular, the convergence of tumor cell and immune cell solutions at $\alpha = 0.8$ is evident for the three fractional-order derivatives.

Furthermore, we analyze the dynamics of tumor cells, hunting immune cells, and resting immune cells under different fractional-order derivatives of the Caputo-Fabrizio fractional derivative. The results indicate that, for smaller fractional orders, the oscillations of tumor cells decrease significantly over a shorter period compared to higher-order fractional derivatives. This phenomenon is illustrated in **Figure 3(a)-(c)**. The reduction of oscillations in a shorter time frame facilitates the scheduling of treatment strategies, while the higher oscillations in tumor growth complicate treatment decisions. This presents a clear advantage over integer order derivatives, as the non-singularity of the kernel aids in studying dynamics as t approaches s.

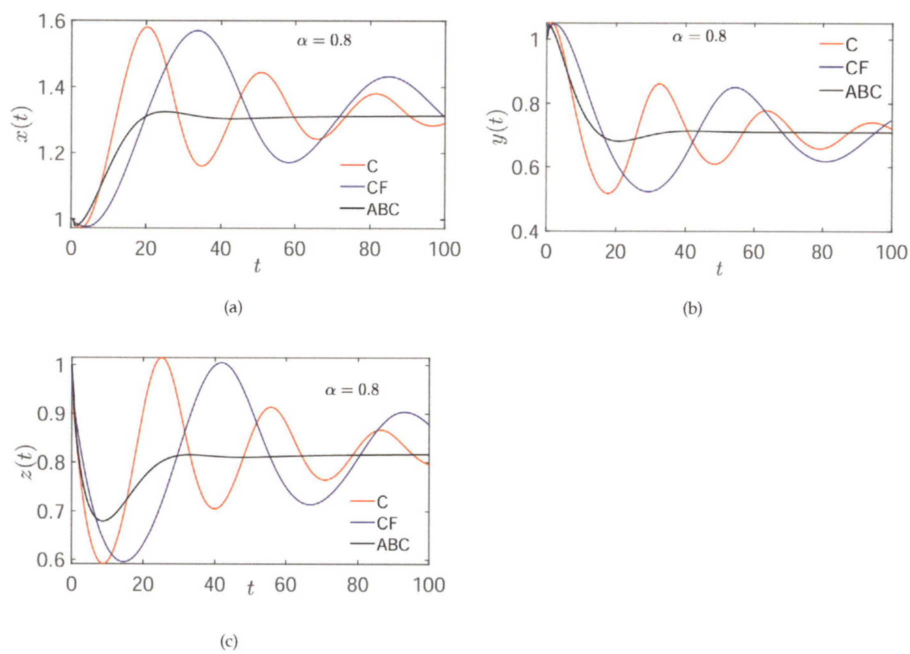

Figure 2.
Comparative view of Caputo-Fabrizio fractional derivative with Caputo and Atangana Beleanu fractional-order derivative.

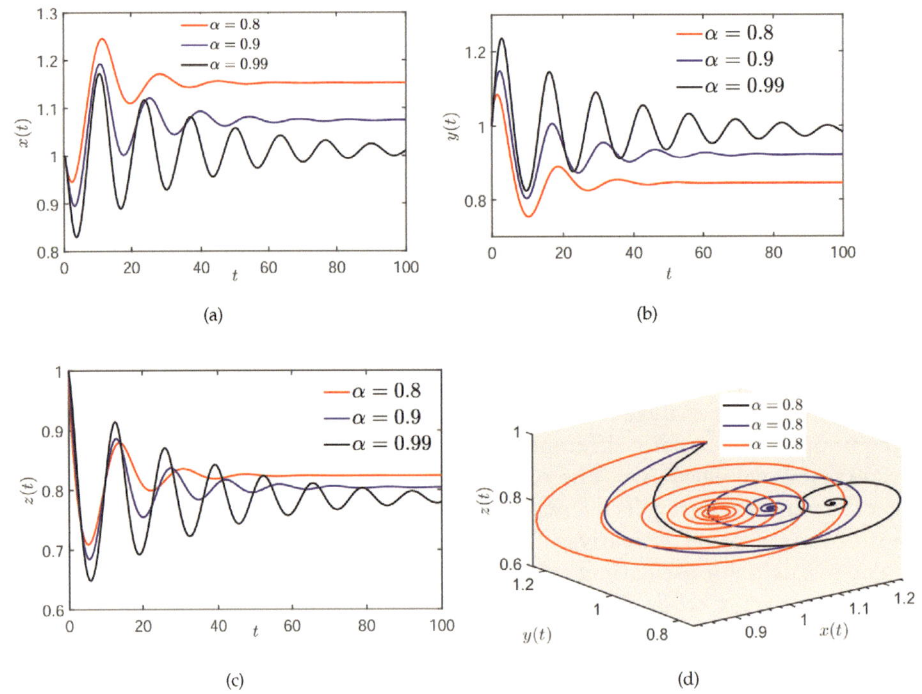

Figure 3.
The dynamics with Caputo-Fabrizio fractional derivatives with different fractional order α.

11. Conclusion

The Riemann-Liouville (RL) type and the Caputo type are two forms of fractional differential operators studied in the field of fractional calculus. However, after a thorough review of the literature, we discovered that these two derivatives are related to a single integral operator. Nevertheless, the fractional integral was found to meet the basic requirements. The RL derivative was utilized to derive the fractional integral in order to prove the foundational theorem of calculus. Atangana recently argued that two types of fractional calculus should be considered: the first being the Caputo type and the second being the RL type, both of which were derived differently. This distinction is important because it is not always possible for the foundational theorem to be satisfied between the Caputo type of derivative and the Riemann-Liouville fractional integral. The Stieltjes integral, which arises in differential equation models of higher-order elasticity and viscoelasticity, is the foundation of the concept of convolution. The time shift, denoted as ρ, represents a summation of elementary relational processes induced by unit step functions and evenly spaced time intervals, in contrast to convolution itself. We can differentiate between the Riemann-Liouville (RL) family of fractional derivatives and the Caputo family of fractional derivatives based on the established principles. Firstly, the Caputo derivative behaves differently from the RL derivative when applied to any function, as the Caputo derivative incorporates integral features, whereas the RL derivative is characterized by its derivative properties. Although there are two types of derivatives—the Caputo derivative and the RL derivative—each has distinct applications in the literature. Angle [47] have demonstrated that the kernel used in the Caputo-Fabrizio (CF) derivative has a cross-over influence on both the probability distribution and the mean square displacement [48]. As a result, the function associated with memory must be specified in accordance with the implication that there is a fixed period $t_1 = a$ such that, until that time, the CF portrays ordinary diffusion, and after that time, it depicts restricted diffusion.

The Caputo-Fabrizio fractional derivative is characterized as a non-local derivative with a non-singular kernel. Its application in mathematical models of cancer has been extensively investigated, underscoring its significance in this field. Comparative studies have been conducted among the Caputo, Caputo-Fabrizio, and Atangana fractional derivatives, offering insights into their respective behaviors and applications. Furthermore, the existence and uniqueness of solutions involving the Caputo-Fabrizio fractional derivative have been established. Stability analyses and various numerical schemes have also been explored, contributing to a deeper understanding of its properties and implications in mathematical modeling. Overall, the Caputo-Fabrizio fractional derivative represents a significant advancement in fractional calculus, particularly in its application to complex biological systems.

Author details

Tarekegn Dinku Gelu[1]*, Boka Kumsa[2] and Jyotrimoy Rana[3]

1 Department of Mathematics, Dambi Dollo University, Ethiopia

2 Department of Mathematics, Wollega University, Ethiopia

3 Department of Mathematics, Indian Institute of Technology Hyderabad, India

*Address all correspondence to: tarekegn.d13@gmail.com

References

[1] Samko SG, Kilbas AA, Marichev OI. Fractional Integrals and Derivatives, Theory and Applications. Amsterdam: Gordon and Breach; 1993. [Engl. Transl. from Russian, Integrals and Derivatives of Fractional Order and Some of Their Applications, Nauka i Tekhnika, Minsk 1987]

[2] Podlubny I. Fractional Differential Equations. San Diego, California, USA: Academic Press; 1998

[3] Monje C, Chen Y, Vinagre B, Xue D, Feliu-Batlle V. Fractional Order Systems and Controls: Fundamentals and Applications. New York: Springer Science & Business Media; 2010

[4] Kilbasi AA, Saigo M. On Mittag-Leffler type function, fractional calculas operators and solutions of integral equations. Integral Transforms and Special Functions. 1996;**4**(4): 355-370

[5] Podlubny I. Fractional Differential Equations. San Diego: Academic Press; 1999

[6] Hu N, Burgueño R. Buckling-induced smart applications: Recent advances and trends. Smart Materials and Structures. 2015;**24**(6):063001

[7] Sabatier J. Fractional dynamical behaviour modelling using convolution models with non-singular rational kernels: Some extensions in the complex domain. Fractal and Fractional. 2025;**9** (2):79

[8] Saxena H. On literature and tools in fractional calculus and applications to mathematical modeling. International Research Journal of Modernization in Engineering Technology and Science. 2021;**3**(12):1014-1019

[9] Baleanu D et al. Fractional Calculus: Models and Numerical Methods. Vol. 3. World Scientific; 2012

[10] Machado JT, Kiryakova V, Mainardi F. Recent history of fractional calculus. Communications in Nonlinear Science and Numerical Simulation. 2011;**16**(3): 1140-1153

[11] Mainardi F, Gorenflo R. Fractional Calculus and Special Functions. Lecture Notes on Mathematical Physics. Bologna, Italy: University of Bologna; 2013. pp. 1-64

[12] Podlubny I. Fractional Differential Equations: An Introduction to Fractional Derivatives, Fractional Differential Equations, to Methods of their Solution and some of their Applications, Mathematics in Science and Engineering. Vol. 198. San Diego, CA: Academic Press, Inc.; 1999. p. MR1658022

[13] Caputo M, Fabrizio M. A new definition of fractional derivative without singular kernel. Progress in Fractional Differentiation & Applications. 2015;**1**:73-85

[14] Atangana A, Baleanu D. New fractional derivatives with non-local and non-singular kernel: Theory and application to heat transfer model. Thermal Science. 2016;**20**:763-769

[15] Salem HAH. Nonlinear differential equations with Caputo fractional derivatives in the class of monotonic functions. Computational Mathematics and Modeling. 2008;**19**(4):430-435

[16] Aydogan MS et al. On high order fractional integro-differential equations including the Caputo–Fabrizio derivative. Boundary Value Problems. 2018;**2018**(1):1-15

[17] Losada J, Nieto JJ. Properties of a new fractional derivative without singular kernel. Progress in Fractional Differentiation & Applications. 2015;**1** (2):87-92

[18] Avalos-Ruiz LF et al. On the dynamics of fractional maps with power-law, exponential decay and Mittag–Leffler memory. Chaos, Solitons & Fractals. 2019;**127**:364-388

[19] Atangana A, Secer A. A note on fractional order derivatives and table of fractional derivatives of some special functions. In: Abstract and Applied Analysis. Vol. 2013, No. 1, p. 279681. Hindawi Publishing Corporation; 2013

[20] Atangana A, Akgül A, Owolabi KM. Analysis of fractal fractional differential equations. Alexandria Engineering Journal. 2020;**59**(3):1117-1134

[21] Jarad F, Abdeljawad T, Baleanu D. On the generalized fractional derivatives and their Caputo modification. 2017

[22] Baleanu D, Mousalou A, Rezapour S. On the existence of solutions for some infinite coefficient-symmetric Caputo-Fabrizio fractional integro-differential equations. Boundary Value Problems. 2017;**2017**:1-9

[23] Miller KS, Ross B. An Introduction to the Fractional Calculus and Fractional Differential Equations. Bristol, England: Wiley; 1993

[24] Abdeljawad T. On Riemann and Caputo fractional differences. Computers & Mathematics with Applications. 2011;**62**(3):1602-1611

[25] Al-Refai M, Pal K. New aspects of Caputo–Fabri IO fractional derivative. Progress in Fractional Differentiation & Applications. 2019;**5**(2):157-166

[26] Ouyang Y, Wang W. Comparison of definition of several fractional derivatives. In: 2016 International Conference on Education, Management and Computer Science. Atlantis Press; 2016

[27] Khurshaid A, Khurshaid H. Comparative analysis and definitions of fractional derivatives. Journal ISSN 2766. 2023:2276

[28] Eftekhari, L, Hosseinpour S, Khalighi M. A comparison between Caputo and Caputo–Fabrizio fractional derivatives for modelling Lotka–Voltera differential equations. Preprint, submitted to Elsevier. 2019

[29] Caputo MR, Ling C. The intrinsic comparative dynamics of locally differentiable feedback Nash equilibria of autonomous and exponentially discounted infinite horizon differential games. Journal of Economic Dynamics and Control. 2013;**37**:1982-1994

[30] Sene N. Stability analysis of the fractional differential equations with the Caputo-Fabrizio fractional derivative. Journal of Fractional Calculus and Applications. 2020;**11**(2):160-172

[31] Marasi HR, Soltani Joujehi A, Aydi H. An extension of the Picard theorem to fractional differential equations with a Caputo-Fabrizio derivative. Journal of Function Spaces. 2021;**2021**(1):6624861

[32] Akman T, Yıldız B, Baleanu D. New discretization of Caputo–Fabrizio derivative. Computational and Applied Mathematics. 2018;**37**:3307-3333. DOI: 10.1007/s40314-017-0514-1

[33] Baleanu D, Mohammadi H, Rezapour S. A mathematical theoretical study of a particular system of Caputo–Fabrizio fractional differential equations for the rubella disease model. Advances

in Difference Equations. 2020;**2020**:184. DOI: 10.1186/s13662-020-02614-z

[34] Toh YT, Phang C, Loh JR. New predictor-corrector scheme for solving nonlinear differential equations with Caputo-Fabrizio operator. Mathematical Methods in the Applied Sciences. 2019; **42**(1):175-185

[35] Matignon D. Stability Results for Fractional Differential Equations with Applications to Control Processing. 1996. pp. 963-968

[36] Li H, Cheng J, Li H-B, Zhong S-M. Stability analysis of a fractional-order linear system described by the caputo-fabrizio derivative. Mathematics. 2019;**7**: 200

[37] Khalighi M et al. Three-species Lotka-Volterra model with respect to Caputo and Caputo-Fabrizio fractional operators. Symmetry. 2021;**13**(3):368

[38] Oldham KB, Spanier J. The Fractional Calculus; Theory and Applications of Differentiation and Integration to Arbitrary Order. San Diego, California, USA: Academic Press; 1974

[39] Ross B. Fractional Calculus and Its Applications. Berlin Heidelberg, Germany: Springer-Verlag; 1975

[40] Li Z, Liu Z, Khan MA. Fractional investigation of bank data with fractal-fractional Caputo derivative. Chaos, Solitons & Fractals. 2020;**131**:109528

[41] Khan MA et al. A fractional order pine wilt disease model with Caputo–Fabrizio derivative. Advances in Difference Equations. 2018;**2018**(1):1-18

[42] Wang JR, Fec M, Zhou Y. Nonexistence of periodic solutions and asymptotically periodic solutions for fractional differential equations. Communications in Nonlinear Science and Numerical Simulation. 2013;**18**(2): 246-256

[43] Alzabut JO, Nieto JJ, GTr Stamov. Existence and exponential stability of positive almost periodic solutions for a model of hematopoiesis. Boundary Value Problems. 2009;**2009**:1-10

[44] Jr V, Carlos A, et al. Can fractional calculus help improve tumor growth models? Journal of Computational and Applied Mathematics. 2020;**379**:112964

[45] Kuang Y, Nagy JD, Eikenberry SE. Introduction to Mathematical Oncology. London: Chapman and Hall/CRC; 2018

[46] Anderson B, Jackson J, Sitharam M. Descartes' rule of signs revisited. The American Mathematical Monthly. 1998; **105**(5):447-451

[47] Tateishi AA, Ribeiro HV, Lenzi EK. The role of fractional time-derivative operators on anomalous diffusion. Frontiers in Physics. 2017;**5**:52

[48] Atangana A. Extension of rate of change concept: From local to nonlocal operators with applications. Results in Physics. 2020;**19**:103515

Chapter 7

Predicting the Behaviour of Oscillators Using Describing Functions

Owen Casha

Abstract

This chapter explores the application of describing functions in predicting the behaviour of non-linear oscillators. Describing functions offer an approximate method that simplifies the analysis of complex, non-linear systems by linearising their frequency response characteristics. This chapter delves into the theoretical foundations of describing functions and their practical use in predicting limit cycles in oscillators, highlighting their capacity to estimate the amplitude and the frequency of the oscillations. Through analytical derivations and case studies, this work demonstrates how describing functions bridge the gap between rigorous non-linear analysis and practical engineering approximations. The limitations, accuracy considerations, and alternative techniques are also discussed, providing a comprehensive understanding of when and how to effectively use describing functions in oscillator analysis and design. The discussion is supported by a literature survey on the use of describing functions in oscillator design and analysis and its relevance to automatic amplitude control mechanisms and the minimisation of AM-to-PM noise conversion.

Keywords: describing functions, non-linear oscillators, limit cycles, automatic amplitude control, AM-to-PM conversion

1. Introduction

The expanding market for communication systems has led to a rising demand for diverse radio frequency (RF) transceivers, driving significant growth in research and development in this field. A key requirement for RF transceiver applications is an integrated local oscillator (LO) that meets specific design criteria, including low phase noise, minimal power consumption, cost-effectiveness, high circuit integration with a small silicon footprint, and a wide tuning range [1]. Predicting the behaviour of a LO is a fundamental component in carrying out its analysis, design and implementation. A practical oscillator is classified as a non-linear system since an ordinary linear differential equation or a transfer function cannot solely model it. An oscillator features intentional non-linear mechanisms that are included on purpose in its feedback loop to provide sustained and predictable oscillations. In addition, an oscillator may also have secondary inherent non-linearities due to both active components and tuning components used in the circuitry. Secondary non-linearities are often undesirable and lead to issues such as amplitude-to-phase (AM-to-PM) noise conversion [2].

A non-linear system can exhibit different modes of equilibrium, some of which are stable and others which are not, depend on the triggering input and initial conditions. Furthermore, a non-linear system such as an oscillator may, under certain conditions, exhibit oscillatory behaviour with a constant amplitude and frequency. This behaviour is called a limit cycle. Limit cycles can be stable or unstable. A stable limit cycle can be triggered even if the input and circuit conditions are not exactly equal to the specific values that correspond to the limit cycle. On the other hand, in the case of an unstable limit cycle, to stimulate it, conditions of the triggering input and circuitry must be perfectly equal to some specific values that depend on the governing equations. A small deviation from these values will fail to trigger the limit cycle. Thus, one must design an oscillator that features a robust and stable limit cycle across a range of process, supply voltage and temperature variations [3].

As yet there is no general or unified tool for analysing non-linear systems of any type. There are several methods, each of which could be applied to a specific type of non-linear system. The phase-plane method is a graphical method, which is accurate but is restricted to second-order systems [4]. The Lyapunov method is the most complete that, in principle, could be used to analyse the stability of any system. However, it involves determining the so-called Lyapunov function for each system to be analysed, and this is not always an easy task [4]. The describing function method (DFM) approximates non-linear elements in a system by an equivalent linear one. While this introduces some errors, it makes the analysis simpler. It handles the problem from a frequency response point of view. The good features of the DFM are that it is a relatively simple technique, it is valid for any system order and it provides information about the system's stability properties, including the presence of limit cycles. Nonetheless, it is based on an approximation: it applies only to non-linear systems whose linear components have low-pass or band-pass filtering properties (thus attenuating high-frequency components) and gives no information about the transient response [4]. Since the DFM provides information on the presence of limit cycles and oscillators include passive networks that exhibit low-pass or band-pass filtering properties, this method is convenient and applicable in predicting the behaviour of oscillators.

The DFM is intuitive and computationally efficient, making it suitable for a preliminary design and insight into oscillation conditions. However, the DFM assumes sinusoidal inputs and neglects higher harmonics, which limits accuracy, especially for strongly non-linear or discontinuous systems. Another technique reported in the literature is called harmonic balance, which, in contrast, includes multiple harmonics in its solution by expressing the oscillator behaviour as a truncated Fourier series and solving for balance in the frequency domain. This method captures non-linearities more accurately than the DFM and can model periodic solutions with greater fidelity. However, it requires solving a non-linear algebraic system, making it computationally heavier than the DFM and less suitable for systems with broadband or chaotic behaviour [5]. Another approach is the numerical continuation method, which traces the evolution of solutions (including steady-state and periodic orbits) as the parameters change, providing powerful insights into bifurcations and stability. Unlike the DFM and harmonic balance, numerical continuation does not rely on simplifying assumptions about the signal shape, making it highly accurate. It can handle complex, multistable, and discontinuous systems. However, it demands significant computational resources and careful initialisation, often requiring prior knowledge of a solution branch [6]. In summary, the DFM is fast and useful for basic predictions, harmonic balance improves accuracy with moderate complexity, and numerical continuation offers high fidelity at the cost of computational effort. The choice depends on the oscillator's non-linearity, required accuracy, and available resources.

In addition to the introductory section, this chapter is divided as follows. **Section 2** presents a concise but informative literature review of various publications which focused on the application of the DFM to analyse and predict the behaviour of oscillators. **Section 3** presents the theoretical background of the DFM. The discussion is aided by applying this approach to analyse the Van der Pol oscillator [4], which is classical example of a non-linear system exhibiting a limit cycle. This section delves into the theoretical foundations of describing functions and their practical use in predicting limit cycles in oscillators, highlighting their capacity to estimate the amplitude and frequency of their oscillations. **Section 4** provides analytical derivations of two oscillator case studies, showing how describing functions bridge the gap between a rigorous non-linear analysis and practical engineering approximations. **Section 5** presents and discusses the relevance of describing functions to automatic amplitude control mechanisms and the minimisation of AM-to-PM noise conversion. Finally, **Section 6** provides some concluding remarks and highlights the main achievements and the limitations of this work while proposing potential future research avenues that may be considered in the coming years.

2. Literature review

2.1 Historical perspective

The describing function (DF) analysis is a classical method for handling non-linear systems and predicting oscillations and limit cycles. Its roots trace back to the 1930s with the work of Krylov and Bogolyubov, who developed a harmonic linearisation approach to approximate non-linear oscillations [7]. The technique was later extended and formalised in the late 1940s and 1950s. In the west, Tustin (UK) and Kochenburger (USA) were early pioneers. The work presented in Ref. [8] applied frequency-response methods to on-off contactor servomechanisms, effectively introducing the DFM to control engineering. In parallel, Soviet researchers like Goldfarb built on Krylov-Bogolyubov's work and made the DFM a standard analysis tool in the USSR by the late 1940s [9]. These developments ran in parallel across countries, with similar concepts emerging in Germany (Oppelt) and France (Dutilh) as well [9]. By the 1960s, the DFM was well established. The work presented in Refs. [10–12] documented the method and broadened its scope to multiple-input systems [12]. The technique's use expanded beyond its control theory origins to oscillator design in electronics. The work [13] showed that describing-function techniques were successfully applied to thermally self-excited mechanical oscillators and were able to predict amplitude and frequency in accurate agreement with experimental investigation [13]. Throughout the following decades, the DFM continued to be used and refined in different fields. In electrical engineering, DFM became an important tool to analyse sinusoidal oscillator start-up and amplitude stabilisation, complementing the linear Barkhausen criterion [11, 13]. Other research also introduced extensions like two-port describing functions to model active devices in oscillators and higher-order describing functions for considering multiple harmonics [12]. These extended forms paved the way for applying DF analysis to modern RF oscillators and systems with more complex non-linearities. Interest in the DF approach also saw a revival in specialised areas, including power electronics circuits such as the self-resonant inverter oscillators, robotics, and biological oscillators [14].

2.2 Strengths and limitations of describing functions

2.2.1 Strengths and advantages

The DF analysis offers several key advantages when dealing with non-linear oscillators. Foremost, it greatly simplifies non-linear system analysis by reducing a non-linear element to an amplitude-dependent transfer function. This quasi-linearisation approach retains essential non-linear behaviour (such as amplitude saturation) while allowing the use of linear analysis techniques, including the Nyquist or the Routh criterion [4, 14]. The DFM can predict the existence and characteristics of limit cycles in a feedback system by treating the non-linear block as if it were a linear gain and phase block that depends on signal amplitude [10]. This enables one to determine the steady-state oscillation amplitude and frequency without the need to solve the full non-linear differential equations. The DF provides a way to find the oscillation amplitude and frequency by solving an algebraic condition at certain amplitude rather than requiring a time-domain simulation [10]. In an operational transconductance amplifier (OTA)-based RC oscillator study, the DFM accurately predicted the stable oscillation amplitude, and the results were confirmed using MATLAB and PSpice simulations [10].

Such agreement confirms that the DF gives meaningful quantitative insight into oscillator behaviour. Another strength is the physical insight it provides into non-linear oscillator dynamics, such as amplitude stabilisation. Traditional linear oscillator theory, such as Barkhausen's criterion, explains how an oscillation starts but not how it limits in amplitude. The describing function fills this gap by modelling how an amplifier's effective gain is effectively adapted to stabilise the amplitude [11]. This explains that a non-linear oscillator requires a form of gain compression (or expansion) to achieve a steady-state response [12]. Indeed, the describing function-based method enhances the oscillator design by complementing Barkhausen's criterion with amplitude control analysis [12]. The method yields intuitive graphical results in the complex plane, where the intersection of the Nyquist plot with the inverse describing function makes it easier to visualise how a limit cycle is sustained [15].

Furthermore, the DF approach is quite generic and widely applicable. It does not depend on the physical nature of the system—only on the presence of a non-linear feedback loop. It has been applied successfully to electronic amplifiers, mechanical structures, and even abstract models, indicating the robustness of the approach [13]. As long as the system can be seen as a linear element in feedback with a single non-linearity, the same DF procedure can predict the oscillation conditions. This cross-domain flexibility implies that such a technique can be applied to both an analogue transistor oscillator and a thermo-mechanical oscillator. The main advantage of the DFM is that it allows the inclusion of the non-linear behaviour of a system while maintaining the simplicity often associated with linear system analysis [14]. It provides a straightforward way to estimate the oscillation amplitude, frequency, and even distortion level [14], which would otherwise require a much more complex analysis.

2.2.2 Limitations and caveats

Despite its usefulness, the DFM has important limitations. It relies on several restrictive assumptions about the system, which, if violated, can cause large errors or analysis failure. The classical DF approach assumes only a single time-invariant memoryless non-linear element exists in the loop and that the oscillation is dominated by its fundamental harmonic [10]. This means the method works best when the non-linearity is a simple static function (such as saturation, cubic gain or dead-zone) and when the oscillator's waveform is nearly sinusoidal. Under those conditions, one can ignore higher harmonics and treat the non-linear block like an effective gain at the fundamental frequency. An additional common assumption is that the non-linearity is odd-symmetric, so it does not produce even-harmonic or d.c. components [10]. While not strictly required, this condition simplifies the analysis by ensuring that the output has no bias and the fundamental frequency is the dominant term. If the system deviates from these assumptions, the DF predictions lose their validity.

One major limitation is the accuracy loss in systems with strong non-linearities or non-sinusoidal oscillations. The DF uses only the first harmonic of the non-linear response. Thus, if the actual oscillator waveform contains substantially higher harmonics, the DF is a crude approximation. In cases where the linear part of the system does not filter out higher order harmonics, such as in relaxation oscillators [16] or digital-like ring oscillators [17], the fundamental approximation breaks down. As a result, the DFM may fail when the higher harmonics significantly influence the behaviour [4]. For bang-bang control systems, the method might predict a limit cycle that is very different from reality or miss an existing oscillation [4]. Similarly, a Schmitt-trigger oscillator, which includes a comparator with hysteresis driving an integrator, produces a highly distorted square or triangle waveform. A basic DF analysis of such a system is likely to be inaccurate or even predict no oscillation since the method cannot capture the piecewise-linear jumps. The output of a Schmitt-trigger relaxation oscillator violates the single-frequency assumption, illustrating where DF analysis is inadequate.

Another limitation is that the DFM requires single-valued non-linearities with no hysteresis or memory. If the non-linearity has memory (such as rate-dependent hysteresis), the simple DFM is not directly applicable [10]. In such cases, more complicated extensions or entirely different methods are needed. Even for static non-linearities, the describing function provides no guarantee of finding all the possible oscillatory solutions—it might find one stable limit cycle, but a system could have multiple limit cycles or complex dynamics exhibiting sub-harmonics or quasi-periodicity. These cannot be predicted by a first-harmonic analysis. Highly non-linear or borderline chaotic systems are essentially beyond the scope of the standard DF approach because they violate the underlying assumption of one dominant frequency. Additionally, the DFM typically gives no information or insight about the transient behaviour or how the oscillation builds up; it only predicts the steady-state amplitude and frequency if a limit cycle exists. It is also worth noting that the DFM, being approximate, sometimes predicts an oscillation that in reality does not occur or misses an existing oscillation. For example, one study applied a two-port describing function to a transistor Colpitts oscillator and found some deviation between the DF-based design predictions and the measured oscillation, which was attributed to modelling errors [12].

This highlights that the DF results are only as good as the model of the non-linear element. If the actual device behaviour differs, such as in the case it has a frequency-dependent non-linearity or significant parasitic dynamics, the DF analysis can be off target. Finally, while graphical DF and Nyquist constructions are convenient, they can become cumbersome for very complex systems, and they do not readily extend to cases with multiple independent non-linearities in the loop. Despite these caveats, many limitations of describing functions are well understood, and there are extensions to mitigate them. For instance, higher-order sinusoidal input describing functions (HOSIDF) may be included to account for second or third harmonics if needed [18]. There are also describing function matrices for multiple-input multiple-output non-linearities and the so-called X-parameters in microwave engineering, which generalise the DF concept to multi-tone steady-state behaviour [12]. These advanced techniques can handle some cases of strong nonlinearity at the cost of more complexity. In summary, the strength of the DF method is in analysing well-behaved single-frequency oscillations with mild non-linearities. When used with awareness of its assumptions and possibly augmented by higher-harmonic corrections, it remains a powerful tool. But for systems that lie outside its valid assumptions, such as those featuring very sharp non-linearities, multi-frequency oscillations, or chaotic regimes, one must resort to time-domain simulation or more exact analytic methods instead of relying on the DFM alone.

2.3 Application in specific oscillator types

2.3.1 Sinusoidal and relaxation electronic oscillators

In electronic oscillator design, DF is often used to design and predict the performance of sinusoidal oscillators that incorporate an amplitude-limiting non-linearity. A classic example is the Wien bridge oscillator. This oscillator uses an amplifier with an automatically adjusting gain (implemented as a small incandescent lamp or a pair of diodes in the feedback path) to stabilise the output amplitude. The describing function of the non-linear lamp or the diode network can be used to find the steady-state oscillation amplitude: essentially, one finds the output voltage at which the non-linear element's effective gain causes the loop gain to be exactly unity. The DFM predicts, for example, the amplitude at which an op-amp Wien bridge will level off given the diode's transfer characteristic curve, and it can estimate the distortion based on how non-sinusoidal the waveform becomes when the limiter engages [14].

Another common application is in LC and RC oscillators using active devices like transistors or operational amplifiers with non-linear gain characteristics [14]. The work presented in Ref. [12] applied a two-port DFM to a BJT Colpitts oscillator design. A describing function for the transistor was derived by modelling it as non-linear two-port admittance and solved for the oscillation in the frequency domain. The design achieved oscillation in the MHz range, and Meszaros et al. [12] reported good agreement between the DF-based analytical predictions and SPICE circuit simulations, as well as reasonable correspondence with experimental measurements [12]. Minor discrepancies were noted, likely due to the transistor model limitations, but these were to be addressed with refined modelling in a follow-up work [12]. The work in Ref. [12] demonstrates how DFM can be used in practical RF oscillator design. By capturing the amplitude-dependent negative resistance of the active device, one can predict the steady-state oscillation amplitude and frequency without a full time-domain simulation. DF analysis has also been used for ring oscillators and relaxation oscillators, although these push the method to its limits. A ring oscillator, which consists of a chain of inverters and produces a square wave, violates the pure single-frequency assumption, yet the work in Ref. [15] has adapted the DF analysis to this special case. The work in Ref. [15] showed that even for a ring oscillator, one can apply a quasi-linear describing function approach by accounting for the phase shift of each inverter stage and the saturation non-linearity of its output, although higher harmonics are significant. In such cases, the describing function is used to estimate the fundamental oscillation frequency and ensure start-up while acknowledging that the waveform is far from sinusoidal. Relaxation oscillators, including multivibrators or the Schmitt-trigger and integrator configuration, are similarly challenging. The DFM can predict the existence of a limit cycle in a Schmitt-trigger oscillator by modelling the hysteresis as an effective non-linear element; indeed, this topology has been used as an example in DF analysis discussions. However, because the oscillation is a highly non-sinusoidal sawtooth or square waveform, a basic DF analysis might give only a rough estimate of frequency and an overly optimistic assumption of a sine-wave output. In practice, one may use DF in relaxation oscillator design simply to verify that a limit cycle will occur under certain bias conditions and then use exact calculations or simulations for the precise waveform and period.

Another important category is automatic gain control (AGC) oscillators and amplitude-regulated oscillators. Many RF and analogue oscillators include an AGC loop or a non-linear gain element to control amplitude, such as a field-effect transistor operated in its non-linear region or a gain-control element driven by a rectified output. The DFM is extremely useful here: it can model the AGC's effect and predict the final amplitude. The work in Ref. [11] emphasises that relying on Barkhausen's criterion alone is inadequate, and one must incorporate non-linear gain compression to truly design an oscillator. DF, coupled with the Routh-Hurwitz stability analysis, forms a systematic method to design oscillators beyond the small-signal threshold point [11]. For instance, in operational amplifier-based oscillators (such as the Wien bridge or phase-shift oscillator), Barkhausen gives the frequency of oscillation, but the DF analysis yields the amplitude at which the non-linearity (or an explicit limiter) of the operational amplifier balances the loop gain to unity [11].

2.3.2 Mechanical and MEMS oscillators

DFM has also been applied to oscillators outside the purely electrical realm, including macroscopic mechanical oscillators and microelectromechanical systems (MEMS) oscillators. The underlying principle—a non-linear feedback loop reaching a self-sustained oscillation—is common to many physical systems, so the DFM can often be ported from circuits to mechanics with appropriate modelling. One of the earliest cross-domain applications was thermally self-excited mechanical oscillators. The work reported in Ref. [13] used DF analysis to predict the behaviour of a pendulum oscillating due to the periodic heating and cooling of a wire, which is a form of self-excited vibration. In Ref. [13], the non-linear relationship between the wire's temperature (and length) and the oscillation amplitude was modelled, and then treated the mechanical system's dynamics as the linear part of a feedback loop [13]. This work showed the broad utility of DF techniques to explain phenomena behind a thermal swing pendulum or a hammock oscillator driven by thermal expansion cycles [13]. While the analysis was approximate, it captured the essential self-limiting nature of the oscillation. The work in Ref. [13] also highlighted that the same DF approach could apply to very different physical energy sources, as long as one can identify a non-linear restoring force or gain element. Beyond thermal oscillators, describing functions have been used to study other mechanical limit cycles. For example, systems with dry friction or backlash can exhibit sustained oscillations (stick-slip motion), and DF analysis has been one of the tools to predict the amplitude of limit cycles in such systems by approximating the friction force versus velocity as a non-linear function. Another application is vehicle or aircraft autopilot oscillations, often called hunting oscillations. Even though these are electro-mechanical systems, classical analyses used DF to estimate limit cycle conditions when a control system had a saturating actuator or a dead zone.

In the context of MEMS oscillators, DF techniques are likewise applicable, although the literature is not as extensive as for electronic circuits. MEMS oscillators, such as vibrating microbeams or micromechanical resonators with an amplifying circuit, often include non-linear effects like amplitude-frequency coupling (Duffing stiffness) or electronic limiting for amplitude stabilisation [19]. DF analysis has been applied in the design of MEMS frequency sources and sensors. For instance, in the design of a silicon micromachined resonant accelerometer [20], an oscillation loop is formed to keep a micro-resonator vibrating, and an automatic amplitude control is used to stabilise the vibration. The describing function for the non-linear amplifier in the loop was derived to calculate the steady-state oscillation amplitude and ensure it stayed within safe limits [20]. This aided the design of the control electronics so that the MEMS resonator operates in a stable limit cycle rather than ringing up to destructive levels or decaying to zero. Similarly, in the study of coupled MEMS oscillators (for injection locking or synchronisation phenomena), simplified describing function models have been used to reduce the non-linear coupling force to an equivalent gain-phase element, which aid in predicting oscillations and phase locking behaviour without heavy computation [21]. These analyses remain somewhat qualitative, but they extend the intuition of describing functions. Another interesting application in the mechanical realm is in robotics and biomechanics. Limit-cycle oscillators appear in legged locomotion (robotic gait engines) and neuro-inspired oscillators (central pattern generators). Some researchers have employed describing function analysis to design rhythmic motion controllers—essentially treating the non-linear robot actuator dynamics as elements in an oscillator loop and using the DFM to ensure a stable limit cycle with desired amplitude [14]. Even for complex systems that can exhibit chaos, like a robot manipulator with non-linear feedback, the DFM has been used to identify periodic orbits and sometimes even to predict the onset of chaotic motion, even though for chaotic systems, other methods are more suitable [14].

3. Describing function method - theoretical background

The DFM is a widely used analytical tool in non-linear control, particularly for systems exhibiting periodic behaviour such as non-linear oscillators. This method approximates non-linear system dynamics by reducing the non-linearity to an equivalent gain and phase shift when subjected to sinusoidal inputs [4]. By characterising the non-linear element using a describing function, one can predict the conditions under which oscillations occur, offering an effective means of analysing self-sustained oscillators. In oscillator design, the DFM identifies the oscillation amplitude and the frequency by locating the intersection between the non-linear describing function and the linear system's frequency response. The method relies on the assumption that the system predominantly responds at the fundamental frequency, simplifying the analysis of complex non-linear interactions. This assumption, although approximate, is often sufficiently accurate for practical oscillator applications. The DFM enables the exploration of stability and performance trade-offs, providing insights into limit cycles and bifurcation phenomena. The method has limitations, such as reliance on accurate modelling of non-linearities and the inability to predict sub-harmonic or chaotic responses.

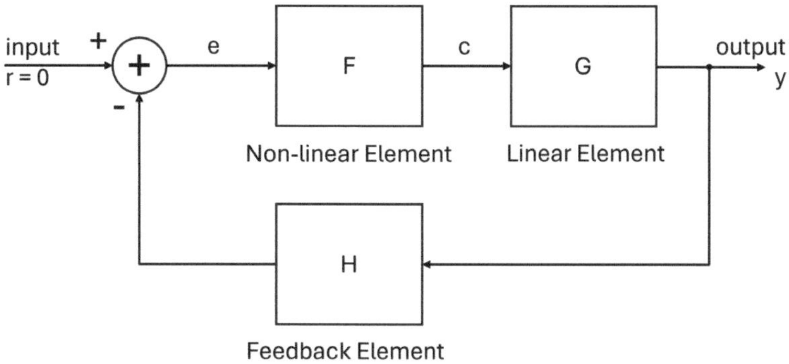

Figure 1.
Block diagram closed loop representation of a generic oscillator.

Figure 1 presents a block diagram representation of a generic oscillator with output y and input r which is set to zero. By means of the DFM, the non-linear element F can be represented by an approximate linear transfer function in terms of the frequency response F_D. Suppose that the input e to the non-linear element F is sinusoidal. The output of the non-linear element c is, in general, not sinusoidal but periodic with the same period as the input e, since the output c would be a complex wave which contains higher harmonics. This output $c(t)$ can be expressed in terms of its Fourier series:

$$c(t) = \frac{A_0}{2} + \sum_{k=1}^{\infty}(A_k \cos(k\omega t) + B_k \sin(k\omega t)) \tag{1}$$

$$A_k = \frac{1}{\pi}\int_0^{2\pi} c(t)\cos(k\omega t)\,d\omega t \tag{2}$$

$$B_k = \frac{1}{\pi}\int_0^{2\pi} c(t)\sin(k\omega t)\,d\omega t \tag{3}$$

where k is the harmonic number and ω is the angular frequency of the non-linear element input signal. In the describing function analysis, one assumes that only the fundamental harmonic component of the output $c(t)$ is significant. Such an assumption is often valid since the higher harmonics in the output of the non-linear element are often of smaller amplitude than those of the fundamental component. In addition, oscillators often include a linear element G that exhibits a low-pass or a band-pass frequency response, with the result that the higher harmonics are highly attenuated with respect to the fundamental harmonic component and so the oscillator output y can be considered as a pure sinusoidal signal, thus confirming the initial assumption that e is sinusoidal.

The describing function of the non-linear element is defined to be the ratio of the phasor representing the fundamental component of the output $c(t)$ to the phasor representing the input $e(t)$:

$$F_D(j\omega, M) = \frac{C_1\angle\Phi_1}{M} \tag{4}$$

where C_1 is the amplitude of the fundamental component of the non-linear element's output, Φ_1 is the phase difference between the output's fundamental component and the input signal $e(t)$ and M is the amplitude of the input signal which is assumed to be sinusoidal. Very often the non-linear elements in oscillators are skew symmetric or odd functions, such that A_0 is zero. Thus, the fundamental component of $c(t)$ would not have any d.c. term and with reference to Eq. (4),

$$C_1 = \sqrt{A_1^2 + B_1^2} \tag{5}$$

$$\Phi_1 = \tan^{-1}\left(\frac{A_1}{B_1}\right) \tag{6}$$

It is thus clear that $F_D(j\omega, M)$ would be a complex quantity when Φ_1 is non-zero. If no energy-storage element is included in the non-linear element, then $F_D(j\omega, M)$ is only a function of the amplitude of the input of the element. This is called a static non-linearity. On the other hand, if an energy-storage element is included, then $F_D(j\omega, M)$ is a function of both the amplitude and frequency of the input. As a summary, since in the context of a non-linear oscillator, there is typically one non-linear element in the loop or various non-linearities that could be grouped as one component and the non-linear element is followed by either a low-pass or band-pass filtering network, one is justified in using the describing function analysis, where one replaces the non-linear block F by $F_D(j\omega, M)$. Thus, one may then apply the classical stability techniques based on the frequency response such as the Nyquist criterion or the Barkhausen criterion.

In order to explore and understand, how the DFM can be applied to analyse the behaviour of non-linear oscillators while identifying its limitations, a classical oscillator known as the Van der Pol oscillator, will be studied and evaluated. The Van der Pol oscillator is a non-linear system described by a second-order differential equation and was introduced by Dutch physicist Balthasar van der Pol in 1927, while working at Philips [4]. The Van der Pol oscillator is governed by Eq. (7):

$$\frac{d^2y}{dt^2} - \alpha\left(1 - y^2\right)\frac{dy}{dt} + y = 0 \tag{7}$$

where y is the output and α is the parameter that controls the non-linearity and the damping. Unlike simple harmonic oscillators, it includes a non-linear damping term (the coefficient of $\frac{dy}{dt}$) that allows for self-sustained oscillations. When y^2 is less than unity, the system is unstable and so y starts to increase due to a negative damping coefficient, while when y^2 is greater than unity, the system start losing energy due to a positive damping coefficient and y starts to decrease. Energy is added or dissipated depending on the system's state, stabilising oscillations at a limit cycle. Rearranging Eq. (7) as follows:

$$\frac{d^2y}{dt^2} - \alpha\frac{dy}{dt} + y = -\alpha y^2\frac{dy}{dt} = \alpha c(t) \tag{8}$$

where

$$c(t) = F(e(t)) = -y^2\frac{dy}{dt} \tag{9}$$

$$e(t) = r(t) - y(t) = -y(t) \tag{10}$$

allows to model the Van der Pol oscillator according to the generic model as shown in **Figure 2**. In this case, the feedback element is equal to unity.

Now as per the DFM, assuming $y(t)$ is a sinusoid with amplitude M:

$$y(t) = Msin(\omega t) \tag{11}$$

gives an expression for the output of the non-linear element $c(t)$:

$$c(t) = -(Msin(\omega t))^2 \frac{d(Msin(\omega t))}{dt} = -(Msin(\omega t))^2 M\omega cos(\omega t) \tag{12}$$

which can be simplified to:

$$c(t) = -\frac{M^3 \omega cos(\omega t)}{4} + \frac{M^3 \omega cos(3\omega t)}{4} \tag{13}$$

by using trigonometric identities. Eq. (13) shows that $c(t)$ is composed of a fundamental frequency equal to that of $y(t)$ and a third harmonic of the same magnitude as the fundamental of $c(t)$. Since the linear element is a second order low-pass filter, the third harmonic of $c(t)$ will be ignored since it will be filtered out. Writing the fundamental of $c(t)$ as:

$$C_1(t) = -\frac{M^2}{4} M\omega \cos(\omega t) = -\frac{M^2}{4}\frac{dy}{dt} \tag{14}$$

the describing function of the non-linear element is obtained as follows:

$$F_D(j\omega, M) = \frac{j\omega M^2}{4} \tag{15}$$

such that the characteristic equation of the Van der Pol oscillator can be expressed as:

$$1 + \frac{\alpha}{(j\omega)^2 - \alpha(j\omega) + 1} \frac{j\omega M^2}{4} = 0 \tag{16}$$

Replacing the non-linear element in **Figure 2** by the describing function given in Eq. (15), one obtains the equivalent linear closed loop model of the oscillator, shown in **Figure 3**.

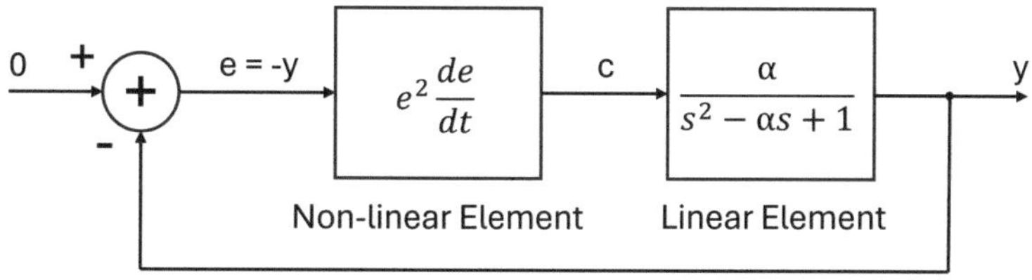

Figure 2.
Non-linear closed loop model of the Van der Pol oscillator.

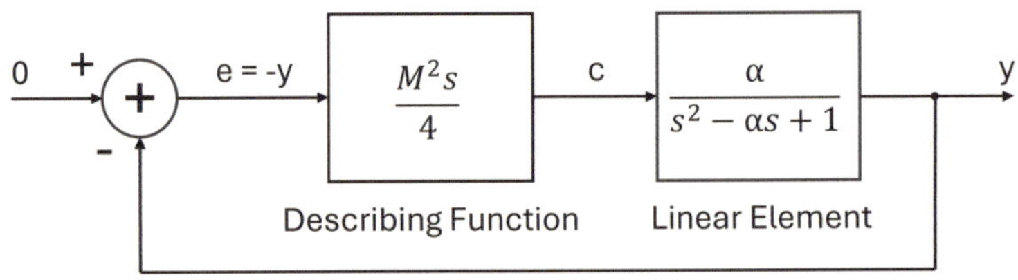

Figure 3.
Equivalent linear s-domain model of the Van der Pol oscillator.

By rationalising Eq. (16), one can express it as a real component:

$$\frac{M^2\alpha^2\omega^2}{(1-\omega^2)^2 + (\alpha\omega)^2} = 4 \tag{17}$$

and an imaginary component:

$$\frac{M^2\alpha(j\omega)(1-\omega^2)}{(1-\omega^2)^2 + (\alpha\omega)^2} = 0 \tag{18}$$

From Eq. (18), one obtains the oscillation angular frequency ω of 1 rad/s, while the oscillation amplitude M of 2 is obtained by substituting $\omega = 1\, rad/s$ in Eq. (17). Eq. (18) provides also another solution ($\omega = 0\, rad/s$), but this is a trivial solution. This result is confirmed by numerically integrating Eq. (7) for a value of $\alpha = 0.1$ and presented in **Figure 4**, where the simulated transient behaviour and the phase portrait of the non-linear oscillator are provided together with the single-sided amplitude spectrum of $y(t)$.

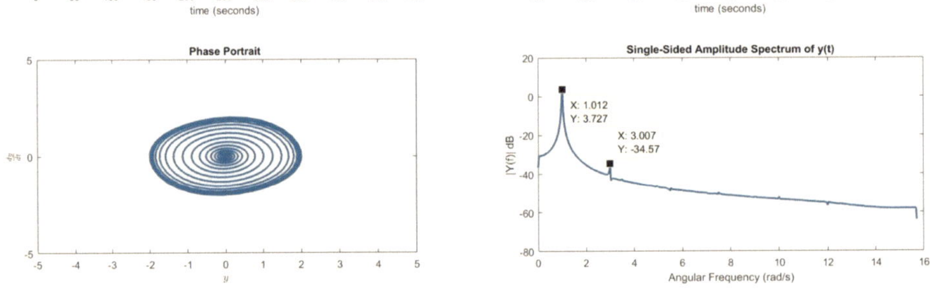

Figure 4.
Simulation results of the Van der Pol oscillator for $\alpha = 0.1$.

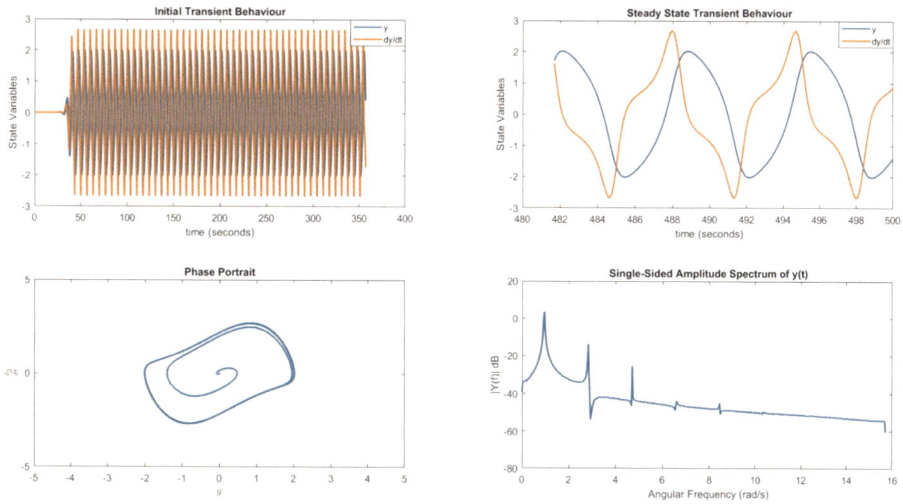

Figure 5.
Simulation results of the Van der Pol oscillator for $\alpha = 1$.

For a low value of α, the DFM is able to predict the behaviour of the non-linear oscillator in terms of amplitude and fundamental oscillation frequency, where the output $y(t)$ is almost purely sinusoidal since the third harmonic is much less than the fundamental (around 38 dB less). The gain frequency response $|G|$ of the linear element is given by Eq. (19):

$$|G| = \frac{\alpha}{\sqrt{(1-\omega^2)^2 + (\alpha\omega)^2}} \tag{19}$$

Eq. (19) shows that when ω is 1 rad/s (fundamental oscillator frequency), $|G|$ is always unity, while the value of $|G|$ at the third harmonic increases with the value of α. This means that in the Van der Pol oscillator, when one increases the value of α, the filtering of the harmonics is less effective, while the gain at the fundamental oscillator frequency remains the same. Thus, the assumption that linear element filters out the harmonics of the output $y(t)$ is less valid when α is increased. This is clearly shown for the simulation of the oscillator when $\alpha = 1$, whose results are presented in **Figure 5**, where the transient behaviour is more non-linear and the oscillator reaches the limit cycle in a shorter time. This shows that in non-linear oscillators, where the filtering action of the linear element is not effective, while the DFM still provides a correct value for the fundamental frequency and amplitude, the accuracy by which the DFM predicts the oscillator behaviour in terms of the total harmonic distortion starts to degrade.

4. Applying the DFM to practical oscillators

This section presents analytical derivations of two oscillator case studies, showing how the DFM can be applied to practical oscillators such as the differential negative resistance oscillator [22, 23] and the Colpitts oscillator [23, 24], to predict their oscillation frequency and amplitude.

4.1 Differential negative resistance oscillator

A circuit that is commonly used in LC tank oscillator designs is shown in **Figure 6** [1]. It uses an NMOS cross-coupled differential pair (M_1 and M_2) to synthesise the negative resistance required to achieve sustained oscillations [1, 22]. Since the outputs of the oscillator ($V_{OUT}+$ and $V_{OUT}-$) are in antiphase and the sources of the transistors are at a small signal differential ground, then the impedance looking down into the drain of M_1 can be shown to be equal to $-1/g_{m1}$, where g_{m1} is the transconductance of M_1.

Thus, to ensure constant amplitude oscillations, this negative resistance must be at least equal to the parasitic resistance of the LC tank connected to M_1. If g_m is too low, any oscillations will decrease in amplitude, while if it is too high, any oscillations will grow exponentially. To achieve constant amplitude oscillations with this oscillator topology, the non-linearity of the transistors is exploited [1, 22]. As the oscillation amplitude increases, the effective g_m of the active devices decreases due to the devices entering the triode region for part of the oscillation period. To ensure the start-up of the oscillator, the transconductance is chosen to produce a start-up gain greater than unity. Then, any initial perturbation grows exponentially, decreasing g_m, until a stable amplitude is reached. It is important to note that this oscillator has two modes of operation. The first mode is referred to as the current-limited regime of operation, in which the tank voltage amplitude is solely determined by the tail-current source I_{BIAS} and the LC tank equivalent resistance. When the amplitude of the tank voltage approaches a voltage greater than the supply V_{DD}, the NMOS transistors enter the linear region at the peaks of the voltage, and so the tank voltage amplitude does not significantly increase beyond that voltage since it is clipped at the source voltage of the NMOS transistors (M_1 and M_2). For the period that the NMOS transistors enter the linear region, they no longer present a negative resistance to the LC tank; thus, the phase noise response is degraded. It is then said that the oscillator has entered the

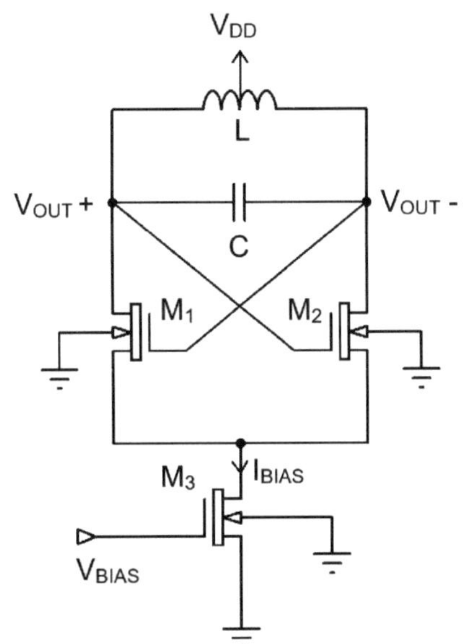

Figure 6.
NMOS-based differential negative resistance LC tank oscillator.

voltage-limited regime of operation. Note also that the tail NMOS transistor M_3 may spend most of its time in the linear region so the tail current does not stay constant but is dependent on the voltage across this NMOS transistor.

This oscillator can be represented using the non-linear closed-loop model given in **Figure 7**, which consists of a non-linear active element, an inversion unity gain that models the inverting nature of the active device, and an LC tank passive equivalent linear element circuit, where C is the capacitance, L is the inductance and R is the total equivalent loss of the tank circuit. The non-linear element is a saturation non-linearity, which is used to model the behaviour of the active device.

For small input signals, the output of the saturation non-linear element is proportional to the input, while for large input signals, the output is constant at the maximum possible output value $\frac{I_{BIAS}}{2}$. Thus for a sinusoidal input given by Eq. (20), the output waveform i_{OUT} becomes as shown in **Figure 8**, where M is the single-ended oscillation amplitude and ω is the angular oscillation frequency.

$$v_{OUT}(t) = -Msin(\omega t) \tag{20}$$

Since the saturation non-linear element is an odd non-linearity, the fundamental Fourier series coefficient A_1 can be shown to be equal to 0 using Eq. (2), while the fundamental Fourier series coefficient B_1 can be computed using Eq. (3). Given the symmetry of the waveform i_{OUT}, B_1 can be expressed as follows:

$$B_1 = \frac{4}{\pi}\int_0^{\frac{\pi}{2}} i_{OUT}(t)\sin(\omega t)d\omega t \tag{21}$$

$$B_1 = \frac{4}{\pi}\int_0^{\omega t_1} g_m M\sin(\omega t)\sin(\omega t)d\omega t + \frac{4}{\pi}\int_{\omega t_1}^{\frac{\pi}{2}} \frac{I_{BIAS}}{2}\sin(\omega t)d\omega t \tag{22}$$

By using Eq. (20) and equating it to $\frac{-I_{BIAS}}{2g_m}$, one may express t_1 as per Eq. (23):

$$t_1 = \frac{\arcsin\left(\frac{I_{BIAS}}{2g_m M}\right)}{\omega} \tag{23}$$

such that

$$B_1 = \frac{2g_m M}{\pi}\left(\arcsin\left(\frac{I_{BIAS}}{2g_m M}\right) + \frac{I_{BIAS}}{2g_m M}\sqrt{1-\left(\frac{I_{BIAS}}{2g_m M}\right)^2}\right) \tag{24}$$

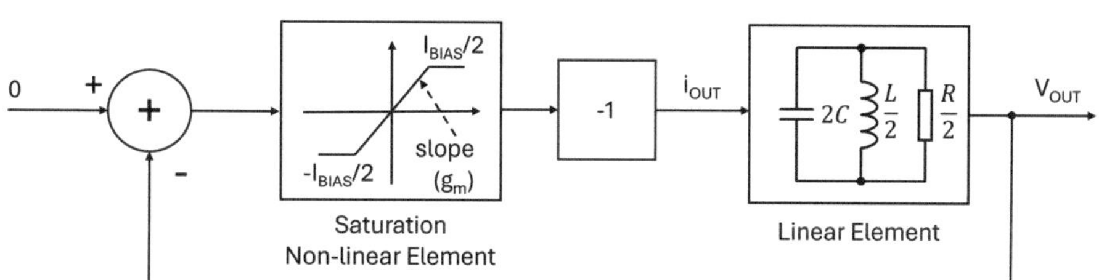

Figure 7.
Single-ended non-linear closed-loop model of the differential negative resistance oscillator.

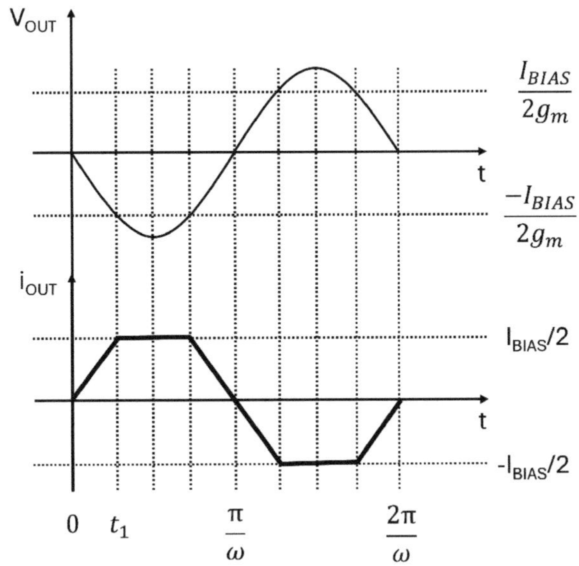

Figure 8.
Input and output waveforms for the saturation non-linear element.

By using Eq. (4), one can show that the static describing function $F_D(M)$ for the saturation non-linearity is given by:

$$F_D(M) = \frac{2g_m}{\pi}\left(\arcsin\left(\frac{I_{BIAS}}{2g_m M}\right) + \frac{I_{BIAS}}{2g_m M}\sqrt{1-\left(\frac{I_{BIAS}}{2g_m M}\right)^2}\right)\angle 0^\circ \tag{25}$$

For a large value of g_m, Eq. (25) can be simplified as follows:

$$F_D(M) = \frac{i_{OUT}}{-v_{OUT}} = \frac{2I_{BIAS}}{\pi M}\angle 0^\circ \tag{26}$$

The transfer function of the linear element $G(j\omega)$ is given by:

$$G(j\omega) = \frac{v_{OUT}}{i_{OUT}} = \frac{\frac{j\omega L}{2}}{(j\omega)^2 LC + \frac{j\omega L}{R} + 1} \tag{27}$$

which can be obtained from the total parallel impedance of the three passive components. By using the Nyquist criterion [4]:

$$-F_D(M)G(j\omega) + 1 = 0 \tag{28}$$

the phase condition of the limit cycle is equal to:

$$\angle F_D(M)G(j\omega) = 0^\circ \tag{29}$$

while the magnitude condition of the limit cycle is equal to:

$$|F_D(M)G(j\omega)| = 1 \tag{30}$$

Since the phase of $F_D(M)$ is equal to 0°, the angular oscillation frequency is thus obtained by using Eq. (29) and finding the frequency at which the phase of $G(j\omega)$ is equal to 0°:

$$\omega = \frac{1}{\sqrt{LC}} \tag{31}$$

while the single-ended oscillation amplitude M is obtained by substituting Eq. (31) into Eq. (30):

$$\frac{2I_{BIAS}}{\pi M}\frac{R}{2} = 1 \tag{32}$$

such that

$$M = \frac{I_{BIAS}R}{\pi} \tag{33}$$

To confirm the validity of this analysis, a differential negative resistance oscillator with an oscillation frequency of around 1.59 GHz, a biasing current I_{BIAS} of 1 mA and a supply voltage V_{DD} of 2.5 V was designed and is shown in **Figure 9**. The NMOS transistors used in the design have a transconductance parameter k_n of $120\mu A/V^2$ and a threshold voltage v_{tn} of 0.6 V. The current mirror biasing transistors M_3 and M_4 have an aspect ratio (W/L) of 25 to achieve an overdrive voltage $(V_{GS} - v_{tn})$ of 0.8165 V, where W is the gate width and L is the gate length of the transistors. The differential pair M_1 and M_2 were designed with an aspect ratio of 200 to achieve a transductance g_m of 4.8 mS and an overdrive voltage of around 0.2 V at a d.c. current of 0.5 mA. As per Eq. (31), the LC tank has a total inductance L of 2 nH and a total capacitance C of 5 pF.

Transient simulations were obtained for the designed oscillator using a generic SPICE simulator for various values of R, which is the total equivalent loss of the LC

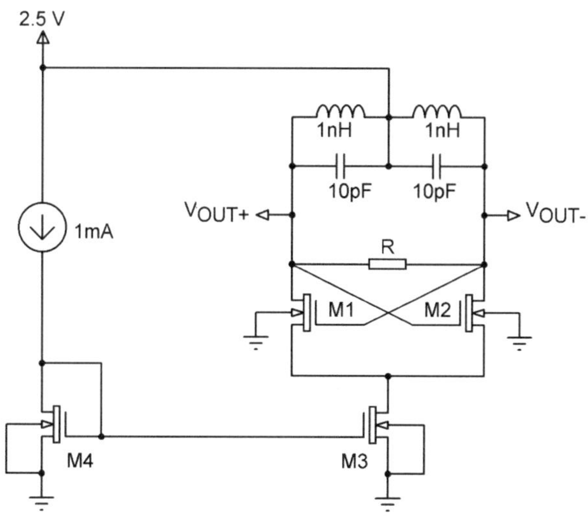

Figure 9.
Schematic diagram of the 1.59 GHz differential negative resistance oscillator.

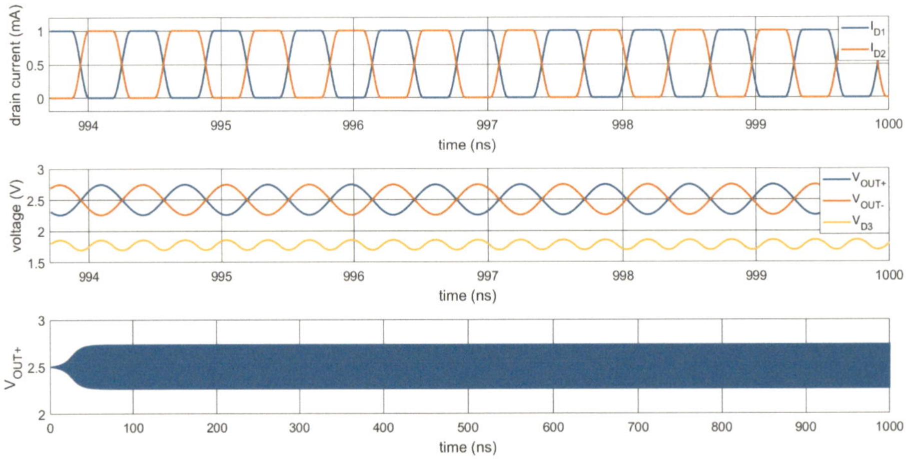

Figure 10.
Transient simulation results of the designed differential negative resistance oscillator with an LC tank loss of 800 Ω *obtained using a generic SPICE simulator.*

tank circuit and mainly depends on the quality factor of the inductor and the capacitor. **Figure 10** shows the steady-state transient simulation results obtained with $R = 800\ \Omega$.

An oscillation amplitude M of 0.243 V over a d.c. voltage of 2.5 V and an oscillation frequency of 1.585 GHz were achieved. **Figure 10** shows the transient plots of the drain currents (I_{D1} and I_{D2}) of transistors M_1 and M_2, the differential voltage outputs V_{OUT+} and V_{OUT-}, the common mode drain voltage of transistor M_3 (V_{D3}) and the oscillation transient built-up. The oscillation amplitude M obtained in the simulation is close to the value predicted by Eq. (33), which is equal to 0.255 V. The discrepancy of around 5% can be attributed to the fact that while the transconductance g_m of transistors M_1 and M_2 is large enough to provide sustained oscillations, it is not infinite and thus, the drain current waveforms are not ideal square waves as assumed in obtaining Eq. (26).

Figure 11 shows a graph that compares the simulated oscillation amplitude with the estimated value given by Eq. (33) for various values of R. For oscillation amplitude

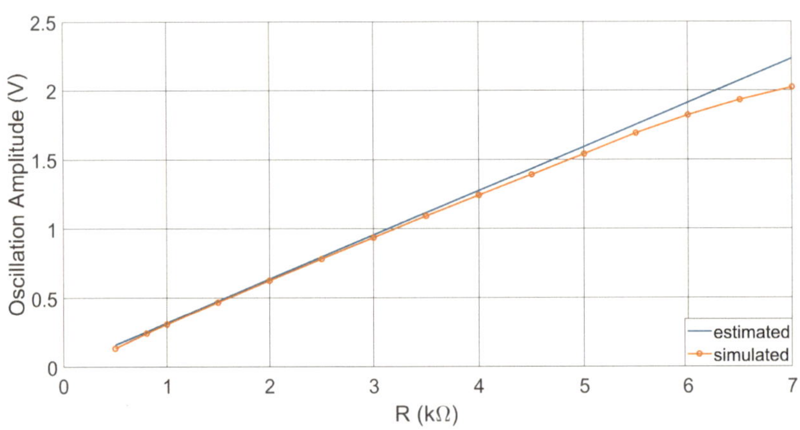

Figure 11.
Comparison of the estimated and simulated oscillation amplitude values for various values of R.

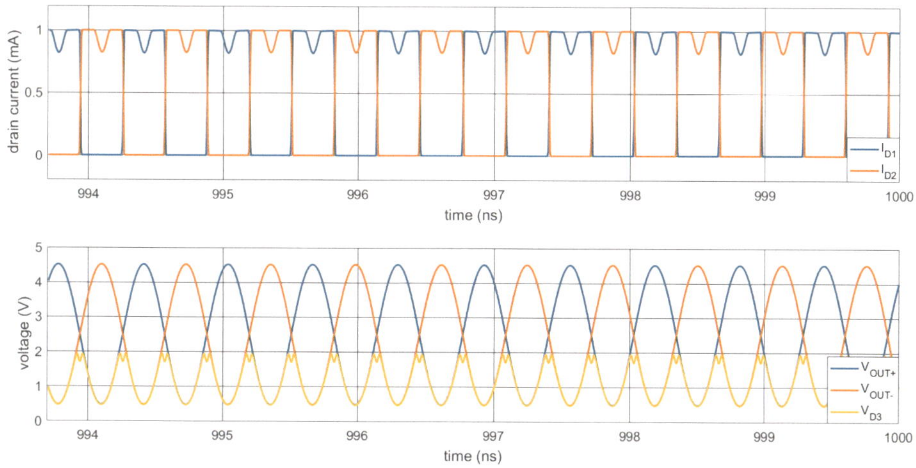

Figure 12.
Transient simulation of the differential negative resistance oscillator with an LC tank loss of 7 kΩ.

values less than 0.935 V (which is equivalent to a R value of 3 $k\Omega$), transistor M_3 always operates in the pinch-off region since its drain-to-source voltage is higher than its overdrive voltage (0.8165 V) and thus, the biasing current I_{BIAS} provided to the cross-coupled differential pair remains constant throughout the oscillation cycle and close to 1 mA.

When M is much greater than 0.935 V, unwanted distortion is introduced as can be noticed from the waveforms of I_{D1}, I_{D2} and V_{D3} in **Figure 12**, since transistor M_3 would not be operating in the pinch-off region during the whole oscillation cycle and a dip is noticed in the biasing current. **Figure 12** presents the transient simulation of the oscillator for R = 7 $k\Omega$. For the given transistor parameters and a d.c. drain current of 0.5 mA, the gate-to-source d.c. voltage of transistor M_1 (V_{GS1}) or M_2 (V_{GS2}) is equal to 0.8 V and the overdrive voltage of transistor M_3 is around 0.8165 V. Thus, if the gate voltage of transistor M_1 (V_{G1}) or M_2 (V_{G2}) is less than 1.6165 V ($V_{DD} - M = 1.565\ V$ where $M = 0.935\ V$), the values provided by Eq. (33) start to become inaccurate. Nonetheless, one would never operate this oscillator beyond this amplitude due to an increase in the total harmonic distortion and AM-to-PM noise upconversion.

4.2 Colpitts oscillator

The Colpitts oscillator is another type of LC tank oscillator that uses a combination of inductors and capacitors to generate a stable sinusoidal output [23]. At its core, it relies on a resonant tank circuit and positive feedback to maintain oscillations without an external input. The circuit's defining feature is the capacitive voltage divider formed by two series capacitors in parallel with an inductor, as shown in **Figure 13**. This arrangement allows a portion of the output signal to be fed back in the correct phase to sustain oscillations. Because of the split-capacitor design, the Colpitts oscillator tends to demonstrate better stability and reduced harmonic distortion compared to alternative oscillators. In transistor-based implementations, one bipolar junction transistor (BJT) or field-effect transistor (FET) is used to amplify the feedback signal and compensate for losses.

Figure 14 shows the equivalent circuit of the LC resonator tank of a Colpitts oscillator where R is the equivalent loss of the tank, L is the inductance and R' is the input resistance seen at the source of the MOS transistor, while C_1 and C_2 form the capacitive divider across the output v_{OUT}. The capacitive divider ratio is given by Eq. (34):

$$\frac{v_F}{v_{OUT}} = \frac{j\omega C_1 R'}{j\omega(C_1 + C_2)R' + 1} \tag{34}$$

which can be approximated by Eq. (35):

$$\frac{v_F}{v_{OUT}} = n = \frac{C_1}{C_1 + C_2} \tag{35}$$

for a high resonant frequency ω. The equivalent admittance of the LC resonator tank can be expressed as Eq. (36):

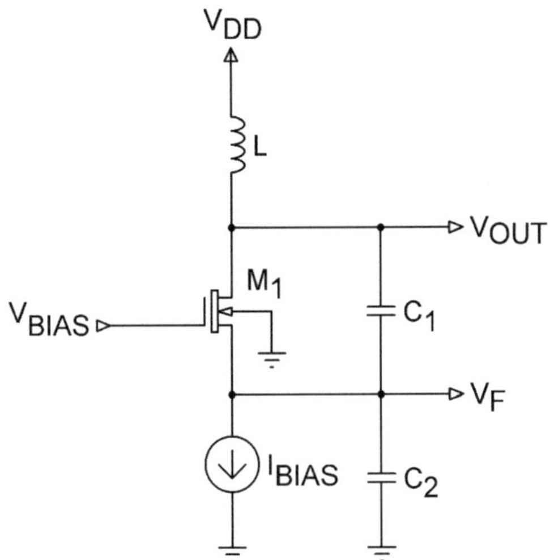

Figure 13.
NMOS-based common-gate Colpitts oscillator (with simplified biasing).

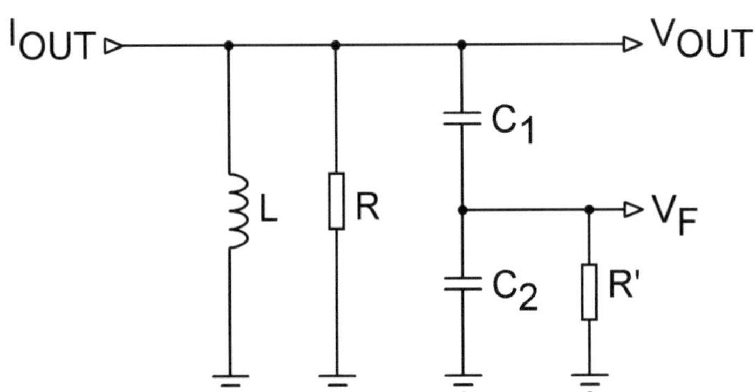

Figure 14.
Equivalent circuit of the LC resonator tank of a Colpitts oscillator.

$$\frac{1}{G(j\omega)} = \frac{i_{OUT}}{v_{OUT}} = j\omega n C_2 + \frac{1}{j\omega L} + \frac{1}{R} + \frac{n^2}{R'} \tag{36}$$

The Colpitts oscillator can be represented using the non-linear closed-loop model given in **Figure 15**, which consists of a non-linear active element, an inversion unity gain that models the inverting nature of the active device, a feedback gain n given by Eq. (35) and an LC tank passive equivalent linear element circuit defined by Eq. (36), where $C_{EQ} = \frac{C_1 C_2}{C_1 + C_2}$ is the equivalent capacitance, L is the inductance and $R_{EQ} = \frac{RR'}{R' + Rn^2}$ is the loaded equivalent loss of the tank circuit, where R' is equal to

$$R' = \frac{1}{g_m} \tag{37}$$

where g_m is the small-signal transconductance of the MOS transistor M_1. **Figure 16** shows the input and output waveforms for the peak detector non-linear element, which is formed by the transistor M_1, which is biased by the current source I_{BIAS}, and the equivalent capacitance C_{EQ}. Close to the negative peak of the output sinusoid v_{OUT}, M_1 injects a current spike of width t_1 and amplitude I_p, which sustains the oscillations.

Using Eq. (2) with $k = 0$, I_{BIAS} can be expressed in terms of I_P:

$$I_{BIAS} = \frac{A_0}{2} = \frac{1}{2\pi} \int_0^{\omega t_1} I_p \, d\omega t = \frac{I_P t_1}{T} \tag{38}$$

where T is the oscillation periodic time, while using Eq. (2) with $k = 1$,

$$A_1 = \frac{1}{\pi} \int_0^{\omega t_1} I_p \cos(\omega t) \, d\omega t = \frac{I_P \sin(\omega t_1)}{\pi} \tag{39}$$

Using Eq. (3) with $k = 1$, gives

$$B_1 = \frac{1}{\pi} \int_0^{\omega t_1} I_p \sin(\omega t) \, d\omega t = \frac{I_P \ (1 - cos\ (\omega t_1))}{\pi} \tag{40}$$

If t_1 is very small (i_{OUT} is ideally a Fermi-Dirac impulse), then A_1 can be expressed as Eq. (41) by using Eq. (38):

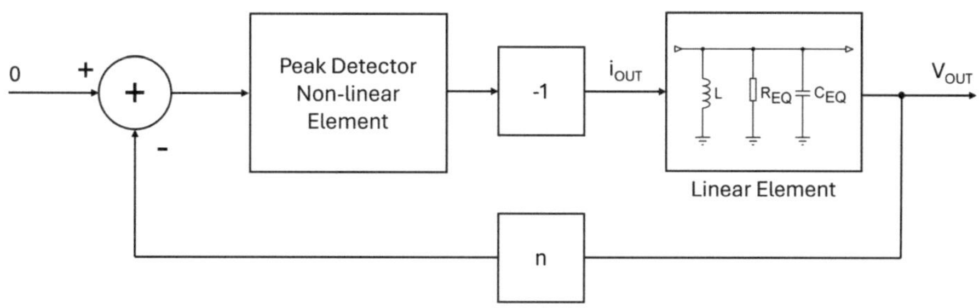

Figure 15.
Non-linear closed-loop model of the Colpitts oscillator.

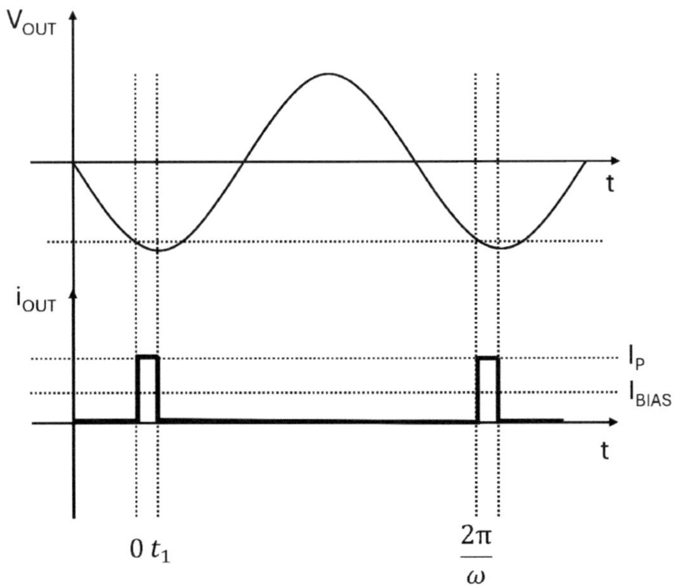

Figure 16.
Input and output waveforms for the peak detector non-linear element.

$$A_1 = \frac{2I_P t_1}{T} \approx 2I_{BIAS} \tag{41}$$

and B_1 is approximately equal to zero. This means that the describing function $F_D(M)$ for the non-linear element in this oscillator can be defined as:

$$F_D(M) = \frac{i_{OUT}}{-v_F} \approx \frac{2I_{BIAS}}{nM} \angle 0^{\circ} \tag{42}$$

where M is the oscillation amplitude. Since the phase of $F_D(M)$ is equal to 0°, the angular oscillation frequency is thus obtained by using Eq. (29) and finding the frequency at which the phase of $G(j\omega)$ is equal to 0°:

$$\omega = \frac{1}{\sqrt{L\left(\frac{C_1 C_2}{C_1 + C_2}\right)}} \tag{43}$$

while the oscillation amplitude M is obtained by substituting Eq. (43) into Eq. (30) and using Eqs. (36) and (42):

$$\frac{2I_{BIAS}}{nM} nR_{EQ} \approx 1 \tag{44}$$

such that

$$M \approx 2I_{BIAS} R_{EQ} \tag{45}$$

To check the validity of this analysis, a common gate Colpitts oscillator with an oscillation frequency of around 60 MHz, a biasing current I_{BIAS} of 1 mA and a supply

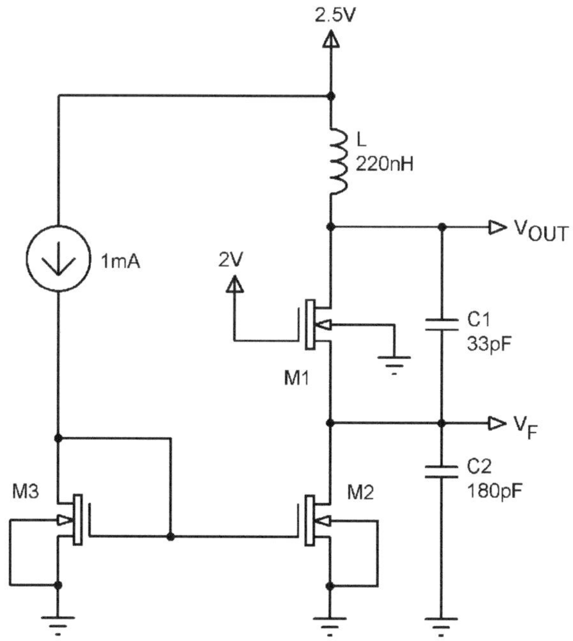

Figure 17.
Schematic diagram of the 60 MHz NMOS based common gate Colpitts oscillator.

voltage V_{DD} of 2.5 V was designed and simulated using SPICE. Its schematic diagram is shown in **Figure 17**. The NMOS transistors used in the design have a transconductance parameter k_n of $120\mu A/V^2$ and a threshold voltage v_{tn} of 0.6 V.

For the particular component values shown, the capacitive divider factor n is estimated to be 0.155 using Eq. (35). Transistor M_1 is biased with a gate voltage V_{G1} of 2 V and overdrive voltage of around 0.2 V at a 1 mA biasing current. The transconductance g_m is thus equal to 9.8 mS such that at resonance, the loaded equivalent loss of the tank circuit R_{EQ} is equal to 708 Ω for R equal to 850 Ω. The absolute minimum acceptable transconductance $g_{m(min)}$ for sustained oscillations is given by Eq. (46) [23] and works out to 8.98 mS:

$$g_{m(min)} \geq \frac{1}{R(n - n^2)} \tag{46}$$

Using Eq. (45), the oscillation amplitude is estimated to be equal to 1.42 V. The SPICE simulation results of the designed oscillator are shown in **Figure 18**, indicating oscillation amplitude of 1.39 V. Eq. (45) is an approximate equation since in the analysis the capacitive divider was treated as an ideal transformer and i_{OUT} was taken to be an ideal Fermi-Dirac impulse. From **Figure 18** one may note that the drain current I_{D1} of transistor M_1 is almost sinusoidal since its g_m is not much larger than the $g_{m(min)}$.

5. Automatic amplitude control and AM-to-PM noise minimization

The DFM discussed in the previous sections provides important insights into the behaviour of oscillators. For the two practical RF oscillators presented and analysed in

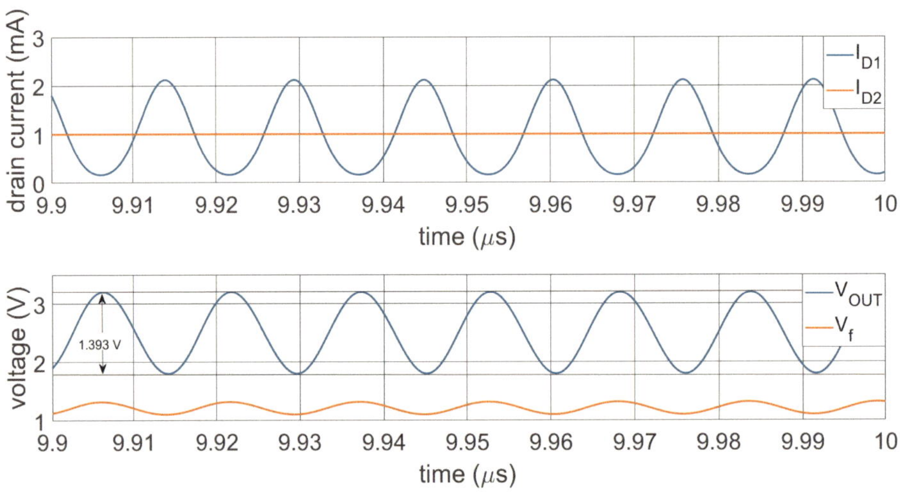

Figure 18.
Transient simulation results of the designed 60 MHz Colpitts oscillator.

Section 4, the DFM shows that they exhibit oscillation amplitude that is independent of the initial conditions present upon powering the circuitry. This is provided through the non-linear active devices used to sustain the oscillations. The DFM models the non-linear active devices by a linearised transfer function $F_D(M)$ as shown by Eqs. (26) and (42). If initially, the oscillation amplitude is small, then $F_D(M)$ is large such that a large current is injected in the LC resonator tank causing the oscillation amplitude to increase since $F_D(M)$ is inversely proportional to M. Eventually, the oscillation amplitude reaches a steady-state value since the effective transconductance $g_{m(eff)}$ of the active devices is reduced to precisely compensate for the loss in the LC resonator tank. For instance, **Figure 19** shows the transient simulation of a differential negative resistance oscillator with $L = 1\,nH$, $C = 1\,pF$, $R = 1\,k\Omega$ and an I_{BIAS} of 2 mA (refer to **Figure 6**). The oscillator exhibits steady-state oscillation amplitude of 0.637 V as per Eq. (33) and an oscillation frequency of 5 GHz as per Eq. (31). While **Figure 19** shows the oscillation build-up, it also shows the transient variation of the effective transconductance $g_{m(eff)}$ given by $|F_D(M)|$, which reaches a steady-state value equal to around 2 mS, providing an open loop gain of unity at the oscillation frequency as per Eq. (30). In fact, the minimum effective transconductance for such oscillator to achieve sustained oscillations given by Eq. (47) [23] confirms this result for R = 1 $k\Omega$.

$$g_{m(eff)} = \frac{2}{R} \tag{47}$$

Eqs. (33) and (45) show that the steady-state value of the oscillation amplitude depends on both the biasing current and the equivalent dynamic resistance of the LC resonator tank. Since this dynamic resistance is a function of frequency, one expects a variation of the oscillation amplitude across the frequency tuning range of the oscillator. Eqs. (33) and (45) highlight the importance of including an automatic amplitude control (AAC) mechanism in an RF oscillator circuit, particularly to one that is designed to provide a wide frequency tuning range [25]. AAC guarantees a reliable oscillator start-up by requiring only a small excess loop gain. This, in turn, helps

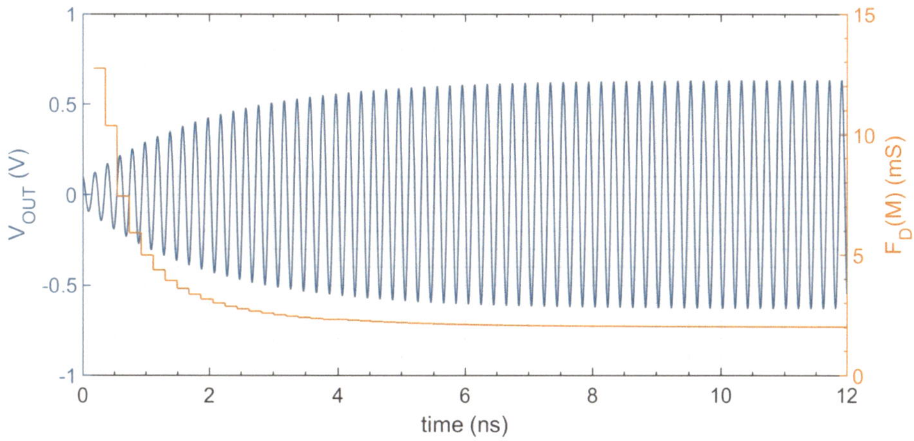

Figure 19.
Transient simulation of a differential negative resistance oscillator.

remove unwanted harmonics from the output and reduces steady-state power consumption thanks to smaller transistor sizes. Moreover, it ensures a constant output power regardless of variations in the resonator quality factor, temperature or process. This is especially crucial for designing the stages that follow the oscillator in frequency synthesiser architectures, as well as for circuits like mixers connected to the oscillator output. Furthermore, it enables setting an optimal bias point to achieve low phase noise [25]. The oscillation amplitude is ideally kept constant over the whole tuning range by varying the biasing current I_{BIAS} so that non-linear effects, such as the modulation of the parasitic and varactor capacitances, which lead to noise upconversion, are minimised [26]. For instance, in the differential negative resistance oscillator discussed in **Section 4.1**, the AM-to-PM conversion of flicker noise from both the biasing (M_3) and negative resistance transistors (M_1 and M_2) can also take place due to the AM-to-PM transfer function exhibited by the cross-coupled transistors [26]. This up-conversion factor depends on both the harmonic distortion at the output nodes and the biasing of the cross-coupled transistor [25, 26].

6. Conclusions and future work

This chapter explored the application of describing functions in predicting the behaviour of non-linear oscillators. It delved into the theoretical foundations of describing functions and their practical use in predicting limit cycles in oscillators, highlighting their ability to estimate the oscillation amplitude and frequency. Through analytical derivations and case studies of some oscillators, this work demonstrated how describing functions bridge the gap between rigorous non-linear analysis and practical engineering approximations. The limitations, accuracy considerations and alternative techniques were also discussed, providing a comprehensive understanding of when and how to effectively use describing functions in oscillator analysis and design. The discussion was supported by a literature survey on the use of describing functions in oscillator design and analysis and its relevance to automatic amplitude control mechanisms and the minimisation of AM-to-PM noise conversion. The following is some future work that one could pursue to advance the field of predicting the behaviour of electronic oscillators using DFM. While some of these may have seen limited consideration in the literature, they remain relatively unexplored or open-ended in terms of achieving comprehensive solutions.

Standard DFM typically use a single-tone sinusoidal input to derive the first harmonic approximation. However, many practical oscillators exhibit higher harmonics, intermodulation products or multi-tone signals in their non-linear operation. Extending DFM to capture multi-frequency inputs or higher-order harmonic content could offer more accurate predictions of phenomena such as distortion, harmonic locking and sub-harmonic or super-harmonic oscillations. Chaos, quasi-periodicity or multi-limit-cycle phenomena can also arise in strongly non-linear circuits. Traditional DFM are aimed primarily at stable limit cycles. Extending DF analyses to detect the onset of chaos or identify quasi-periodic attractors, potentially by examining sub- and super-harmonic components or analysing multiple limit cycles, is an interesting research avenue.

Practical oscillator implementations have time-varying parameters such as temperature-dependent transistor parameters, ageing or bias circuit drift. Describing functions often rely on the assumption of a time-invariant non-linearity. Adapting the DFM theory to time-varying behaviours, such as piecewise or slowly varying modelling, to study how oscillator characteristics evolve is another relevant research question. This might incorporate periodic re-linearisation or perturbation-based techniques. Oscillator performance in practical applications is significantly affected by noise, which degrades phase noise. Traditional describing function-based methods ignore or only marginally address noise. One may formulate a stochastic describing function methodology that captures the interplay between non-linearities and noise sources, potentially coupling DF analyses with the statistical methods or stochastic differential equations.

Multiple oscillators may be coupled together, such as injection-locked oscillators [27], quadrature output oscillators [28] or multi-phase oscillator arrays [29]. Linear or single-device DF analyses may fail to account for coupling. It would be interesting to develop network-level DFM where each oscillator or stage is represented by a describing function, and the interactions among these blocks are modelled to predict collective phenomena like synchronisation, injection locking, mode hopping and pattern formation. By pursuing these directions, one can expand the DFM and apply it to a wider range of practical oscillators, from low-frequency relaxation circuits to high-frequency RF or mm-wave designs, while incorporating emerging complexities such as multi-tone signals, stochastic noise and intricate coupling between multiple non-linear elements.

Dedication

I dedicate this book chapter to my beloved son, Harry, who will always be in my thoughts. May this note bring him solace and happiness, knowing that his mother, Charmaine, and I have loved him as our own child from the moment our paths crossed and eagerly awaited the day he would be in our arms.

Author details

Owen Casha
Department of Microelectronics and Nanoelectronics, University of Malta, Msida, Malta

*Address all correspondence to: owen.casha@um.edu.mt

References

[1] Casha O, Grech I, Micallef J. Issues on the design and implementation of radio frequency CMOS LC tank voltage-controlled oscillators. Journal of Analog Integrated Circuits and Signal Processing. 2009;**61**(2):159-170

[2] Pepe F, Bonfanti A, Levantino A, Samori C, Lacaita AL. A wideband voltage-biased LC oscillator with reduced flicker noise up-conversion. In: Proceedings of the 2013 IEEE Radio Frequency Integrated Circuits Symposium; 2–4 June 2013; Seattle, WA, USA. USA: IEEE; 2013. pp. 27-30

[3] Raab B, Kaufmann M, Rauchenecker M, Ostermann T. Robust and efficient design plan for integrated differential negative-gm LC VCOs. In: Proceedings of the 22nd International Conference Mixed Design of Integrated Circuits and Systems; 25–27 June 2015; Torun, Poland. USA: IEEE; 2015. pp. 386-389

[4] Modern OK. Engineering C. 5th edition. London, UK: Pearson; 2021

[5] Krack M, Gross J. Harmonic Balance for Nonlinear Vibration Problems. Switzerland: Springer Cham; 2019

[6] Allgower EL, Georg K. Numerical Continuation Methods - an Introduction. Heidelberg: Springer Berlin; 1990

[7] Krylov N, Bogolyubov N. Introduction to Nonlinear Mechanics. New Jersey, USA: Princeton University Press; 1947

[8] Kochenburger RJ. A frequency response method for analyzing and synthesizing contactor servomechanisms. Transactions of the American Institute of Electrical Engineers. 1950;**69**(1):270-284

[9] Bissell CAA. Andronov and the development of soviet control Engineering. IEEE Control Systems Magazine. 1998;**18**(1):56-62

[10] Pranayanuntana P, Anuntahirunrat K, Fongsamut C, Kaewsaiha P. The describing function method and the analysis of the magnitude stabilization phenomenon in a nonlinear oscillators. In: Proceedings of IEEE International Symposium on Communications and Information Technology; 12–14 October 2005; Beijing, China. USA: IEEE; 2005. pp. 420-423

[11] Rodríguez-Vázquez A, Leñero-Bardallo JA. When Barkhausen's criterion does not suffice and you must rely on the forgotten art of oscillator design. In: Proceedings of the 2024 XVI Congreso de Tecnología, Aprendizaje y Enseñanza de la Electrónica (TAEE); 26–28 June 2024; Malaga, Spain. USA: IEEE; 2024. pp. 1-6

[12] Meszaros G, Ladvanszky J, Berceli T. Oscillator design using two-port describing functions. In: Proceedings of IEEE International Conference on Computer Communications; 10–15 April 2016. San Francisco, CA, USA. USA: IEEE; 2016. pp. 1-9

[13] Reich HJ. Describing-function analysis of thermally self-excited mechanical oscillators. IEEE Transactions on Education. 1968;**11**(2): 116-119

[14] Vidal E, Poveda A, Ismail M. Describing functions and oscillators. IEEE Circuits and Devices Magazine. 2001;**17**(6):7-11

[15] Wang T. Analyzing oscillators using describing functions. arXiv: 1710.02000. 2017:1-19

[16] Stork M. Voltage controlled oscillators: Sinusoidal and square wave ring and relaxation oscillators. In: Proceedings of the 3rd Mediterranean Conference on Embedded Computing; 15–19 June 2014; Budva, Montenegro. USA: IEEE; 2014. pp. 180-183

[17] Jahangir MZ, Paidimarry CS. Design of a Novel Charge Pump based current starved ring oscillator with reduced phase noise. In: Proceedings of the 2023 International Conference for Advancement in Technology; 24–26 January 2023; Goa, India. USA: IEEE; 2023. pp. 1-3

[18] Eijk LF, Kostić D, Khosravi M, Hosseinia SH. Higher order sinusoidal-input describing function analysis for a class of multiple-input multiple-output convergent systems. IEEE Transactions on Automatic Control. 2025;**70**(1): 673-680

[19] Chen D, Wang Y, Chen X, Huang W, Xie J. Duffing nonlinearity localization via extension energy confinement in an elastic mode semicircular beams resonator. IEEE Electron Device Letters. 2019;**40**(2): 314-317

[20] Sung S, Chang BS, Lee KY, Lee YJ. A vibration-controlled resonant accelerometer design and its application to the single structured gyroscope/ accelerometer system. In: Proceedings of the 2007 IEEE International Conference on Vehicular Electronics and Safety; 13–15 December 2007; Beijing, China. USA: IEEE; 2007. pp. 1-6

[21] Mouro J, Pinto R, Paoletti P, Tiribilli B. Microcantilever: Dynamical response for mass sensing and fluid characterization. Sensors (Basel). 2020; **21**(1):115

[22] Pinto AM, Souza RRN, Solano JEV, Lima ER, Manera LT. LC-VCO design with switched Varactor Array for L-band in 65 nm CMOS technology. In: Proceedings of the 2024 IEEE International Conference on Design, Test and Technology of Integrated Systems; 14–16 October 2024; Aix-EN-PROVENCE, France. USA: IEEE; 2024. pp. 1-6

[23] Lee TH. The Design of CMOS Radio-Frequency Integrated Circuits. 2nd ed. UK: Cambridge University Press; 2003

[24] Chaichana A, Mhuenpong K, Sotner R, Jaikla W. Electronically controllable Colpitts oscillator with amplitude controllability using commercially available ICs. In: Proceedings of the 13th International Conference on Power, Energy and Electrical Engineering; 25–27 February 2023; Tokyo, Japan. USA: IEEE; 2023. pp. 22-26

[25] Casha O, Grech I, Micallef J, Gatt E. Design of a 1.2 V low phase noise 1.6 GHz CMOS buffered quadrature output VCO with automatic amplitude control. In: Proceedings of the 13th IEEE International Conference on Electronics, Circuits and Systems; 10–13 December 2006; Nice, France. USA: IEEE; 2006. pp. 192-195

[26] Casha O, Grech I, Micallef J, Gatt E. Design considerations and device selection in the implementation of low phase noise LC-VCOs. In: Proceedings of the 14th IEEE International Conference on Electronics, Circuits and Systems; 11–14 December 2007; Marrakech, Morocco. USA: IEEE; 2007. pp. 1083-1086

[27] Xin K. Frequency domain modeling and performance analysis of injection-locked LC oscillator. IEEE Transactions on Circuits and Systems II: Express Briefs. 2024;**71**(1):11-15

[28] Kadam M, Kudabay Y, Oliveira P, Wohlmuth HD. Issakov V. A low phase noise 12 GHz LO generation with quadrature outputs in 0.13 μm SiGe BiCMOS. In: Proceedings of the 15th German Microwave Conference; 11–13 March 2024; Duisburg, Germany. USA: IEEE; 2024. pp. 93-96

[29] Dutta S, Ranjan R, Kumari P, Sinha P, Ranjan RK, Singh DK. Operational Transconductance amplifier (OTA) based dual mode multiphase oscillator. In: Proceedings of the 5th International Conference on Recent Trends in Computer Science and Technology; 9–10 April 2024; Jamshedpur, India. USA: IEEE; 2024. pp. 332-337

Chapter 8

Grayscale Image Security Based on Nonlinear 2D Chaotic Map and Bit Pair Rotation

Manzoor Ahmad Lone and Shaima Qureshi

Abstract

In this study, the nonlinear 2D logistic chaotic map is used in an efficient and effective way for the encryption of digital images. In most of encryption algorithms, images are exposed for confusion and diffusion process at pixel level to form a cipher image. In this scheme, the chaotic sequences generated by 2D logistic chaotic map in combination with bit pair rotation are used for confusion and diffusion processes. Original image is disturbed at both pixel level and bit level to form a more reliable and strong cipher. The scheme is verified against the test images taken from the USC-SIPI image database. The simulation results of the quality parameters, favorable outcomes of theoretical value tests, and comparative analysis with the existing encryption schemes verify the reliability and robustness of the proposed encryption approach.

Keywords: security, 2D chaotic logistic map, confusion, diffusion, bit level, pixel level

1. Introduction

With the advancements in multimedia technology and the extensive use of digital devices and the internet, protecting digital images from unauthorized access has become essential. Images, unlike text, have properties such as large data size, high redundancy, and strong pixel correlation, making conventional text encryption unsuitable. To address this challenge, researchers have investigated various image cryptosystems, utilizing techniques including wavelet transforms, chaos theory, cellular automata, DNA coding, and Fourier transforms. Within these, encryption schemes based on chaos theory [1–12] have gained significant attention because of the characteristics, including extreme responsiveness to starting parameters, indeterminacy, ergodicity, and strong randomness. The chaotic streams produced by chaotic maps are highly complex and difficult to analyze or predict, thereby strengthening the security of the cryptosystem. As a result, chaos theory has become an appealing foundation for designing cryptographic systems aimed at addressing information security challenges. In comparison with chaos-based image encryption schemes at the pixel level, relatively less research has focused on bit-level encryption approaches. Bit-level encryption offers notable advantages, as it enhances both confusion and diffusion processes during image encryption. Various bit-level techniques have been proposed [2, 4, 8, 13–18] in the literature to strengthen image security. Zhu et al. [13] explored the 8-bit representation of gray scale pixels, where the four MSBs are permuted individually, while the four LSBs are treated as a group and relocated together to improve efficiency. The permutation process is guided by the Arnold cat map, with its parameters generated using the Logistic map. In the diffusion stage, every pixel element is sequentially modified using both the chaotic iteration values and the outcome of the Logistic map. This approach achieves significant speed and to ensure resilience against differential attacks, a minimum of three encryption rounds is essential. Similarly, Xu et al. [2] proposed a bit-plane encryption scheme that separately partitions pixels into four higher- and four lower-order groups, from which binary streams are obtained, and applies chaos-driven diffusion in combination with cyclic shifts and XOR operations. Thereafter, the binary stream elements are interchanged in accordance with the ordering specified by piecewise linear chaotic maps. Remarkably, these two encryption phases provide strong security with just a single round of processing. In Ref. [4], a novel 22D-LICM is introduced, constructed on the foundation of the CCMC model. The encryption system applies both bit-level permutation and diffusion. In the scheme, the circular shift process performs bit-level permutation and achieves the bit-level diffusion by XOR and reverse operations. The results of the study indicate that the proposed scheme achieves strong encryption performance and considerable efficiency. The authors in Ref. [19] proposed a novel image security scheme that integrates a two-dimensional Hénon Sine Map with DNA encoding techniques. In the method, pixel elements are permuted using the 2D Hénon Sine Map and subsequently diffused through DNA encoding rules. Similarly, the authors in Ref. [20] achieved simultaneous permutation and diffusion by combining chaotic theory with DNA encoding, aiming to ensure secure transmission of grayscale images over public networks. In Ref. [21], authors propose a security scheme that integrates a regulated zigzag transformation coupled with bit-level encryption in a quantum walk framework. In Ref. [22], authors present a cryptographic security evaluation of a color image encryption method that utilizes bit-level operations and an enhanced 1D chaotic map (CIEA-IOCM). In the proposed CIEA-IOCM, pseudo-random streams produced by the enhanced chaotic map are used to perform bit-level permutation-diffusion followed by a linear conversion, thereby converting the plain image into its encrypted counterpart. In Ref. [14], inspired by the efficiency of permutation process by bit-level, the authors proposed a lightweight bit-level confusion scheme combined with cascade cross-circular diffusion to strengthen cryptosystem security and decrease computational redundancy typically observed in the traditional frameworks..

The above studies motivated the authors to propose an image security approach based on the chaotic theory using the 2D logistic map. The 2D logistic map yields two pseudo-random streams that when integrated with bit-pair rotation, enhance confusion and diffusion in the input image, and furnish a secure cipher. To enhance security and enlarge the key space, the initial key values of the 2D logistic map are generated based on the plain image. The algorithm exhibits high sensitivity to even minimal changes (Δ) in its key parameters. Slight modifications in the initial conditions (A_0, B_0) or the control parameters (l_1, l_2, k_1, k_2) trigger a significant avalanche effect in the cipher images, thereby ensuring robust resistance against various cryptanalytic attacks. Moreover, the full set of key parameters in the two-dimensional logistic map generates an extensive key space that effectively safeguards the system from brute-force attacks.

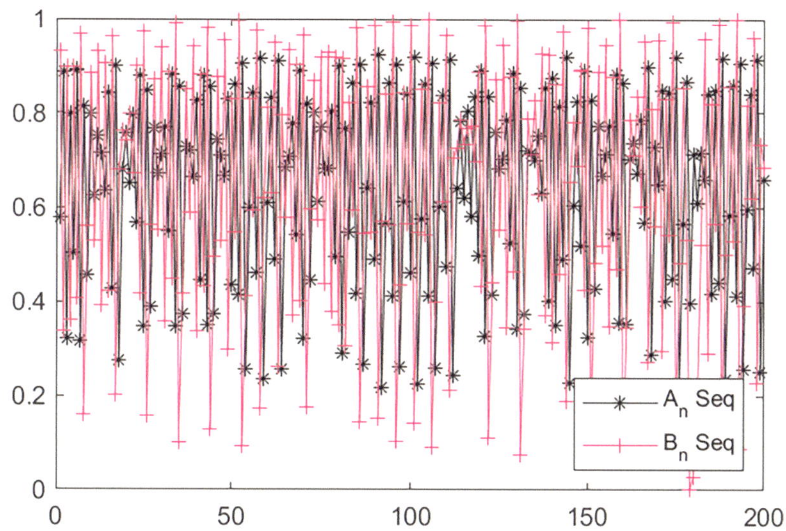

Figure 1.
A_n and B_n chaotic plots of 2D logistic map.

2. 2D logistic map

The 2D logistic map is derived as an extension of the 1D logistic map, which serves as a fundamental nonlinear dynamical model for investigating chaotic phenomena. The 1D logistic map is characterized by a single equation as the 2D logistic map involves two coupled equations, generating two interdependent chaotic streams. The map increases the system's complexity and unpredictability, making it suitable for different cryptographic applications, such as image encryption. The 2D logistic map is represented by the following mathematical formula [23]:

$$A_{n+1} = l_1 A_n (1 - A_n) + k_1 B_n, \tag{1}$$

$$B_{n+1} = l_2 B_n (1 - B_n) + k_2 A_n^2 + A_n B_n \tag{2}$$

When the system parameters ($l_1, l_2, k_1,$ and k_2) are set within the range $2.75 < l_1 \leq 3.4$, $2.75 < l_2 \leq 3.45$, $0.15 < k_1 \leq 0.21$ and $0.13 < k_2 \leq 0.15$ the system exhibits chaotic behavior and produces two chaotic sequences A_n and $B_n \in (0, 1]$. **Figure 1** presents the chaotic sequences A_n and B_n generated by the two-dimensional chaotic map.

3. Proposed encryption method

The steps of the proposed encryption algorithm are given below:

Algorithm 1. The proposed procedure:

Input: Original image (im^o_{MN}), keys.
Output: Cipher image (im^E).

1. Use the input image im^o, and on the similar pattern of [7], obtain the two initial key values of the 2D logistic map as follows:

$$A_0 \leftarrow \frac{1}{256} \mathrm{mod}\left(\sum im^o, 256\right), \tag{3}$$

$$B_0 \leftarrow A_0/2. \tag{4}$$

2. Use Eqs. (1) and (2) to generate A_n and B_n chaotic streams, respectively.

3. Scale the sequences A_n and B_n in the interval of $[0 - 255]$, and then and convert them into MN matrices say A_{MN} and B_{MN}, respectively.

4. XOR A_{MN} and B_{MN} matrices as shown below:

$$Z_{MN} \leftarrow \left(A_{[MN]} \oplus B_{[MN]}\right). \tag{5}$$

5. Form a row vector RV_{1xn} from Z_{MN}, where $n = MN$

6. Employ pseudocode $[RV', indx] \leftarrow \mathbf{Sort}(RV)$; to Sort RV sequence.

7. Shuffle the pixels in im^o by employing pseudo-code $im^{o_1} \leftarrow im^o(indx)$, and then transform im^{o_1} into a matrix say im'_{MN}. Convert im'_{MN} into binary form say im^{bin} and apply bit pair rotation in right direction to all the 8-bit pixels of im^{bin}. The output of this step is a bit pair rotated image say im^{binRot}.

8. Convert im^{binRot} into decimal form say img''.

9. Finally, perform XOR operation between img'' and Z_{MN} as shown follows:

$$im^E \leftarrow (im'' \oplus Z_{MN}). \tag{6}$$

Where im^E represents the final cipher image.

The plain image is reconstructed during the decryption phase by sequentially applying the inverse operations of the encryption process.

4. Result analysis

For simulation purposes, test images of size 256×256 were obtained from the standard USC-SIPI image database [24], as depicted in **Figure 2**. The simulations were performed on a system with IntelR, CoreTM i7 CPU 2.40GHz and 16GB operating on Windows 10 using MATLAB software.

4.1 Key space

The key space characterizes the strength of an encryption scheme in resisting brute-force attacks. To effectively withstand exhaustive search attacks, the key space should be at least ≥ 128-bits [25]. With a computational precision of 10^{-15}, the overall

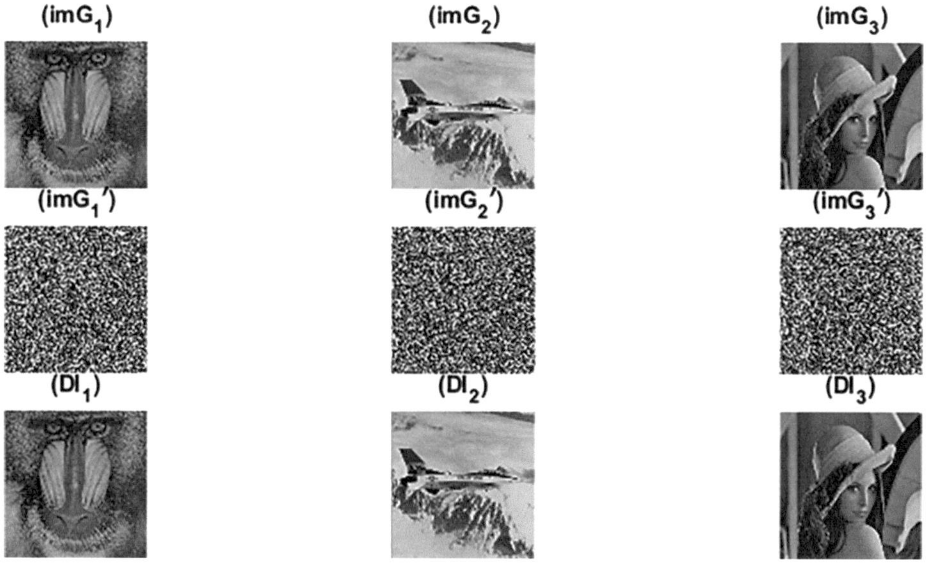

Figure 2.
ImG_1(Baboon), imG_2(Jet), imG_3(Lena).

key space of the proposed scheme is estimated to be 2^{299}. Moreover, in the proposed system, the initial key parameters of the 2D logistic map are dynamically linked to the gray image, contributing to an additional key space of 2^{8MN}. Hence, the proposed encryption framework provides an extensive key space, ensuring strong resistance against brute-force cryptanalytic attacks.

4.2 Key sensitivity

In an effective encryption scheme, minor variations in the secret keys should induce a pronounced avalanche effect in the cipher image, and the decryption process should yield unintelligible images. In the simulation, a slight modification of a single key within the key set produced erroneous images, as illustrated in **Figure 3**. These results demonstrate that the proposed system is highly sensitive to small changes in key parameters.

4.3 Information entropy

This indicator determines the randomness present in an image. The mathematical relation of information entropy is given by the Eq. (7):

$$E(I) = \sum_{a=0}^{T} p(I_a) \log_2 \frac{1}{p(I_a)}, \tag{7}$$

where $T = 2^{n-1}$, and $p(I_a)$ denotes the probability of I_a. The theoretical reference value of $E(I)$ in a 256 gray-scale image is 8 [25]. The numerical results of the proposed encryption system yield entropy values close to this reference, as presented in **Table 1**, thereby confirming the high randomness of the ciphered image data.

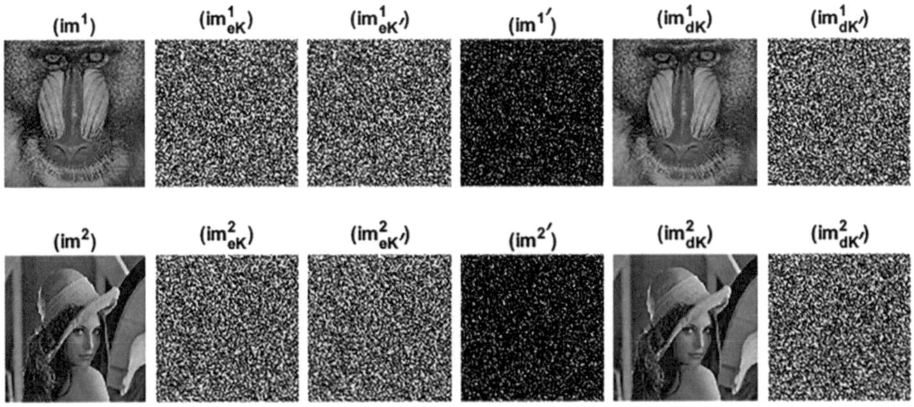

Figure 3.
(im^1, im^2) plain images, (im^1_{eK}, im^2_{eK}) encrypted images by correct key, $(im^1_{eK'}, im^2_{eK'})$ encrypted images by incorrect key, $(im^{1'}, im^{2'})$ difference between correct key and incorrect key encrypted images, (im^1_{dK}, im^2_{dK}) decrypted images of im^1_{eK}, im^2_{eK} by correct key, $(im^1_{dK'}, im^2_{dK'})$ decrypted images of im^1_{eK}, im^2_{eK} by incorrect key.

Img.	$\boldsymbol{E(I)}$
Baboon	7.9974
Jet	7.9976
Lena	7.9973
Avg.	7.9974

Table 1.
Information entropy results.

4.4 Histogram analysis

The histogram of an original image typically exhibits spikes across various intensity levels, reflecting a higher concentration of pixels in those regions. In a secure cipher image, such patterns should be eliminated, resulting in a histogram with a uniform distribution. In the proposed system, the histograms of the cipher images, as illustrated in **Figure 4**, are evenly distributed compared to those of the original images. This demonstrates that the proposed method effectively resists histogram-based attacks.

4.5 Correlation analysis

An effective encryption scheme should eliminate the high correlation typically present between adjacent pixels in an original image [5]. The correlation coefficient $\varphi_{(f,f')}$ is given as follows:

$$\varphi_{(p,p')} = \frac{F \times \mathrm{Q}_{(p,p')}}{\sqrt{\sum_{f=1}^{F}\left(p_f - E_p\right)^2}\,\sqrt{\sum_{f=1}^{F}\left(p'_f - E_{p'}\right)^2}}, \tag{8}$$

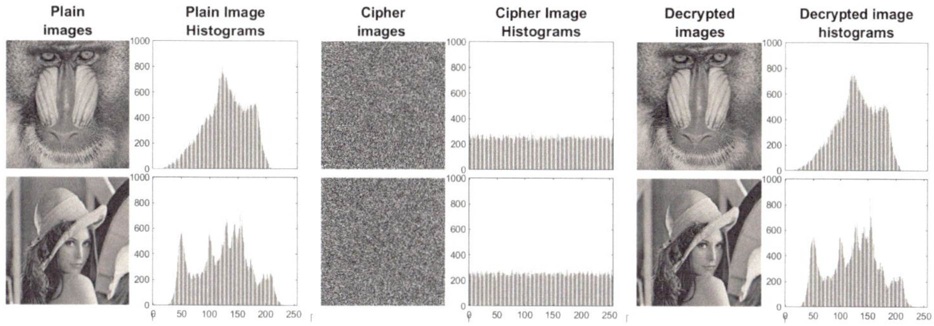

Figure 4.
Histogram plots of plain and encrypted images.

Img.	HC	VC	DC
Baboon	0.0000	0.0026	−0.0023
Jet	−0.0005	−0.0027	0.0050
Lena	−0.0060	0.0008	−0.0039
Avg.	−0.0022	0.0002	−0.0004

Table 2.
Correlation coefficient.

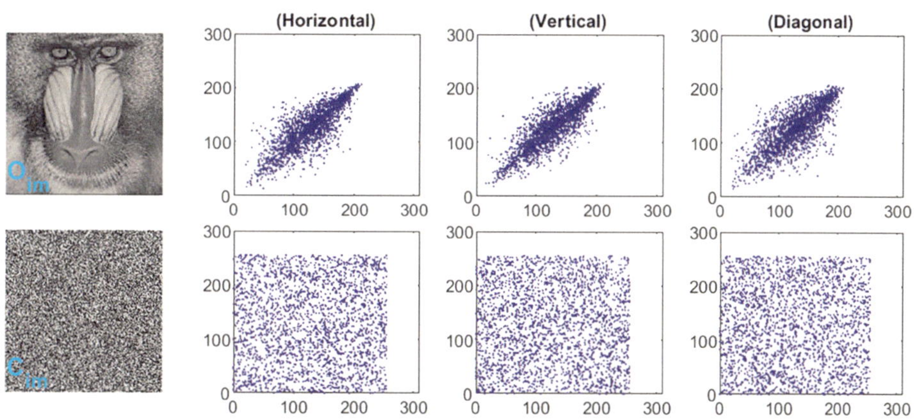

Figure 5.
Correlation plots of original image (O_{im}) *and cipher image* (C_{im}).

where F = size of image, $\varrho_{(p,p')} = E\left(\left(p - E_p\right)\left(p' - E_{p'}\right)\right)$, $E_p = \frac{1}{F}\sum_{f=1}^{F} p_f$, and (p, p') is the adjacent pixel pair. The results are shown in **Table 2** and the plots in **Figure 5** justify that the proposed system can resist the statistically-based attacks.

4.6 Comparison

The comparison analyses of the various parameters of the proposed system shown in **Tables 3–5** are acceptable and are comparable with the existing encryption

Img.	Entropy			
	Ref. [20]	Ref. [26]	Ref. [19]	Prop.
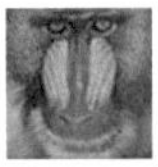	7.9938	7.9975	7.9971	7.9974
	7.9974	7.9972	7.9970	7.9973
	7.9958	7.9970	7.9974	7.9972
Avg.	7.9957	7.9973	7.9972	7.9974

Table 3.
Comparison analysis of entropy with existing schemes.

Img.	Ref. [20]			Ref. [26]			Ref. [19]			Prop.		
	HC	VC	DC	HC	VC	DC	HC	VC	DC	HC	VC	DC
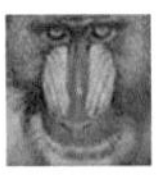	0.0059	0.0041	0.0028	0.0065	0.0337	0.0244	0.0026	0.0009	0.0052	0.0008	−0.0037	0.0025
	0.0062	0.0074	0.0009	0.0229	0.0103	0.0100	0.0028	0.0041	0.0010	−0.0019	−0.0003	0.0059
	0.0037	0.0258	0.0079	0.0211	0.0129	0.0013	0.0016	0.0059	0.0034	0.0007	0.0036	0.0032
Avg.	0.0051	0.0100	0.0029	0.0142	0.0228	0.0086	0.0025	0.0035	0.0029	−0.0002	0.0009	0.0010

Table 4.
Correlation coefficient comparison.

Algorithm	Prop.	Ref. [25]	Ref. [20]	Ref. [27]	Ref. [28]	Ref. [29]
Key space	2^{299}	2^{232}	2^{240}	2^{186}	2^{219}	2^{298}

Table 5.
Key space comparative results.

schemes. Therefore, they justify the performance and effectiveness of the proposed image security scheme.

5. Conclusion

In this study, the nonlinear two-dimensional (2D) logistic chaotic map is effectively employed for digital image encryption. The integration of the 2D logistic map with bit-pair rotation enables effective scrambling and diffusion of image pixels, resulting in a secure cipher image. Two pseudo-random streams generated by the 2D logistic map, in conjunction with bit-pair rotation, are systematically utilized to maximize pixel confusion and diffusion. To further strengthen security and expand the key space, the initial key values of the 2D logistic map are made dependent on the input image, ensuring that even minimal perturbations in the initial conditions result in incorrect cipher images and prevent successful decryption. The robustness of the scheme is validated using standard test images from the USC-SIPI image database. The simulation results, including the evaluations of quality parameters, theoretical benchmark tests, and comparative analyses with existing encryption techniques, demonstrate the effectiveness, reliability, and resilience of the proposed encryption approach.

Author details

Manzoor Ahmad Lone[1*] and Shaima Qureshi[2]

1 Department of CSE, University of Kashmir, NorthCampus (Delina), Baramulla, J&K, India

2 Department of CSE, National Institute of Technology, Srinagar, J&K, India

*Address all correspondence to: mahmadlone@gmail.com

References

[1] Belazi A, Talha M, Kharbech S, Xiang W. Novel medical image encryption scheme based on chaos and DNA encoding. IEEE Access. 2019;**7**:36 667-36 681

[2] Xu L, Li Z, Li J, Hua W. A novel bit-level image encryption algorithm based on chaotic maps. Optics and Lasers in Engineering. 2016;**78**:17-25

[3] Li C, Luo G, Qin K, Li C. An image encryption scheme based on chaotic tent map. Nonlinear Dynamics. 2017;**87**(1): 127-133

[4] Cao C, Sun K, Liu W. A novel bit-level image encryption algorithm based on 2D-LICM hyperchaotic map. Signal Processing. 2018;**143**:122-133

[5] Lone MA, Qureshi S. RGB image encryption based on symmetric keys using Arnold transform, 3D chaotic map and affine hill cipher. Optik. 2022;**260**: 168880

[6] Luo Y, Yu J, Lai W, Liu L. A novel chaotic image encryption algorithm based on improved baker map and logistic map. Multimedia Tools and Applications. 2019;**78**(15):22 023-22 043

[7] Niyat AY, Moattar MH, Torshiz MN. Color image encryption based on hybrid hyper-chaotic system and cellular automata. Optics and Lasers in Engineering. 2017;**90**:225-237

[8] Diaconu A-V. Circular inter–intra pixels bit-level permutation and chaos-based image encryption. Information Sciences. 2016;**355**:314-327

[9] Wu J, Liu Z, Wang J, Hu L, Liu S. A compact image encryption system based on Arnold transformation. Multimedia Tools and Applications. 2021;**80**(2): 2647-2661

[10] Lone MA, Qureshi S. A novel and efficacious use of three-dimensional intertwining logistic map for the security of greyscale images. The Imaging Science Journal. 2021;**69**(5–8):288-301

[11] Zhou Y, Hua Z, Pun C-M, Chen CP. Cascade chaotic system with applications. IEEE Transactions on Cybernetics. 2014;**45**(9):2001-2012

[12] Pellicer-Lostao C, López-Ruiz R. Role of symmetry and geometry in a chaotic pseudo-random bit generator. International Journal of Computer Science and Software Technology. 2008;**4**:43-53

[13] Zhu Z-L, Zhang W, Wong K-W, Yu H. A chaos-based symmetric image encryption scheme using a bit-level permutation. Information Sciences. 2011;**181**(6):1171-1186

[14] Zhang W, Wong K-W, Yu H, Zhu Z-L. An image encryption scheme using lightweight bit-level confusion and cascade cross circular diffusion. Optics Communications. 2012;**285**(9): 2343-2354

[15] Sheng S, Wu X. A novel bit-level image encryption scheme using hyper-chaotic systems. In: 2013 10th International Conference on Fuzzy Systems and Knowledge Discovery (FSKD). Shenyang, China: IEEE; 2013. pp. 1015-1019

[16] Fu C, Huang J-B, Wang N-N, Hou Q-B, Lei W-M. A symmetric chaos-based image cipher with an improved bit-level permutation strategy. Entropy. 2014; **16**(2):770-788

[17] Zhang Y, Xiao D. An image encryption scheme based on rotation

matrix bit-level permutation and block diffusion. Communications in Nonlinear Science and Numerical Simulation. 2014; **19**(1):74-82

[18] Zhang W, Yu H, Zhao Y-L, Zhu Z-L. Image encryption based on three-dimensional bit matrix permutation. Signal Processing. 2016;**118**:36-50

[19] Wu J, Liao X, Yang B. Image encryption using 2D Hénon-sine map and DNA approach. Signal Processing. 2018;**153**:11-23

[20] Enayatifar R, Abdullah AH, Isnin IF, Altameem A, Lee M. Image encryption using a synchronous permutation-diffusion technique. Optics and Lasers in Engineering. 2017;**90**:146-154

[21] Zhang T, Wang S. Image encryption scheme based on a controlled zigzag transform and bit-level encryption under the quantum walk. Frontiers in Physics. 2023;**10**:1097754

[22] Wen H, Chen R, Yang J, Zheng T, Wu J, Lin W, et al. Security analysis of a color image encryption based on bit-level and chaotic map. Multimedia Tools and Applications. 2024;**83**(2):4133-4149

[23] Zhang Q, Liu L, Wei X. Improved algorithm for image encryption based on DNA encoding and multi-chaotic maps. AEU-International Journal of Electronics and Communications. 2014;**68**(3): 186-192

[24] USC-SIPI Image Database. Available from: https://sipi.usc.edu/database/

[25] Ayubi P, Setayeshi S, Rahmani AM. Deterministic chaos game: A new fractal based pseudo-random number generator and its cryptographic application. Journal of Information Security and Applications. 2020;**52**:102472

[26] Hosny KM, Kamal ST, Darwish MM, Papakostas GA. New image encryption algorithm using hyperchaotic system and Fibonacci Q-matrix. Electronics. 2021; **10**(9):1066

[27] Farah M, Farah A, Farah T. An image encryption scheme based on a new hybrid chaotic map and optimized substitution box. Nonlinear Dynamics. 2020;**99**(4):3041-3064

[28] Hu X, Wei L, Chen W, Chen Q, Guo Y. Color image encryption algorithm based on dynamic chaos and matrix convolution. IEEE Access. 2020; **8**:12 452-12 466

[29] Wu J, Cao X, Liu X, Ma L, Xiong J. Image encryption using the random FRDCT and the chaos-based game of life. Journal of Modern Optics. 2019; **66**(7):764-775

Chapter 9

The Importance of Establishing the Continuity of Investment in Infrastructure: A Perspective Based on Structural Analysis

Byron J. Idrovo-Aguirre, Héctor Acuña and Pablo Easton

Abstract

This chapter formally demonstrates that ensuring the continuous development of infrastructure investment plans helps mitigate inefficiencies in capital formation caused by bureaucratic constraints. This issue is particularly significant because creating a structural or long-term investment planning mechanism that remains independent of economic and political cycles requires more precise interventions than those necessary for the broader modernization of the state. To frame the discussion, we develop a structural model of dynamic optimization for infrastructure investment, grounded in firm theory with a rational expectations environment, adjustment costs in capital stock formation, and financial constraints. The model parameters have been calibrated specifically for the Chilean economy.

Keywords: dynamic programming, capital adjustment costs, infrastructure investment, Bayesian estimation, calibration, structural modeling

1. Introduction

The timely completion of infrastructure projects significantly enhances quality of life, either by providing prompt access to services offered by new capital stock or by boosting employment through increased productive capacity in firms that benefit from the new infrastructure. However, large-scale infrastructure projects in Chile—such as high-complexity hospitals, irrigation works, bridges, tunnels, and road networks—take, on average, over 12 years to materialize. This is roughly double the average duration observed in OECD countries, according to the National Evaluation and Productivity Commission (CNEP) [1]. The inefficiencies in infrastructure investment are closely tied to the lack of modernization within the state apparatus [2]. For instance, bureaucratic barriers, the absence of explicit mechanisms for public works procurement, and issues in risk-sharing arrangements between public clients and contractors in response to unexpected service changes all serve to hinder the sector's efficiency and development. Notably, Refs. [1, 2] indicate that 65% of the inefficiencies in infrastructure investment can be attributed to the lack of state modernization, while the remaining 35% of the productivity gap is linked to factors managed by firms within the industry, as documented by Refs. [1–4].

In this context, the present research formally demonstrates that instituting the continuous development of infrastructure investment plans—specifically, plans that are largely independent of economic and political cycles—can significantly counteract the inefficiencies in capital stock formation brought about by systemic bureaucracy. Our findings indicate that the creation of an Infrastructure Agency or a similar mechanism tasked with planning, prioritizing, and managing sectoral investments could reduce the average time to complete major infrastructure projects by between 15 and 30%. This would imply a reduction from an average of 12 years to a range of 6 to 10 years for the completion of such projects. The relevance of this result lies in the fact that creating a structural investment planning mechanism requires more targeted actions than the broader, more complex challenge of modernizing the state as a whole. This conclusion is supported by a structural model of dynamic optimization for infrastructure investment, calibrated specifically for the Chilean economy and assuming an imperfect market environment, drawing on the works of Love [5] and by Gilchrist and Himmelberg [6]. In particular, Love [5] provides evidence that financial development influences growth by mitigating financial constraints that would otherwise distort the efficient allocation of investments. For her analysis, the author estimates a semi-structural investment model incorporating financial leverage rigidities. While Love [5] research question lies beyond the scope of our study, her model—which closely follows the work of Gilchrist and Himmelberg [6]—proves valuable for our purposes.

The model assumes that entrepreneurs seek to maximize the present value of their firm, represented as the discounted value of expected dividends, subject to constraints related to capital accumulation and external financing. Specifically, the equilibrium investment equation is derived by rearranging the first-order conditions for the optimization problem. This equation provides the intertemporal decision rule governing capital stock accumulation across consecutive periods [7–10]. Thus, the optimal level of investment is determined by equating the marginal cost of postponing investment decisions with the marginal benefit of investing. The key modifications to the models of Love [5] and Gilchrist and Himmelberg [6] are as follows: (i) Most of the parameters are calibrated using official data and specialized literature on Chile, while other parameters are estimated using Bayesian techniques; (ii) the variables of the model are normalized by the GDP of infrastructure construction, as an approximation of macroeconomic theory regarding balanced growth paths [11].

The document is organized as follows: The first section serves as the introduction. The second section details the structural model of sectoral investment and its optimal equilibrium conditions, based on firm theory incorporating market imperfections. The third section presents the parameter values of the model, combining calibration methods with Bayesian estimation techniques. The fourth section introduces the simulation scenarios. Finally, the fifth section offers concluding remarks.

2. The model

Consider an economy consisting of a continuum of construction firms engaged in civil works and non-residential building projects, operating within an environment characterized by market imperfections. These imperfections are modeled through financial rigidities and adjustment costs related to capital stock formation—essentially representing frictions to investment. Consequently, in this economy, firms seek to maximize their profits while contending with challenges such as limited access to credit and inefficiencies in the investment process. Examples that justify modeling the adjustment costs in capital formation include the implicit costs stemming from disputes between clients and contractors, delays in project approval within the environmental impact assessment system, and idiosyncratic factors (i.e., inefficiencies specific to individual organizations). As a result, the process of capital stock formation (involving infrastructure such as roads, bridges, dams, ports, and hospitals) takes longer than what would be predicted by models assuming complete markets.

Let $V_t(k_t, \xi_t))$ represent the expected future profits of a representative construction firm in the sector, expressed in present value terms. Thus, the firm aims to solve the following optimization problem:

$$V_t(k_t, \xi_t) = \max_{\{i_{t+s}\}_{s=0}^{+\infty}} E_t \sum_{s=0}^{+\infty} \beta^s d_{t+s} \quad (1)$$

Subject to:

$$d_t = \pi(k_t, \xi_t) - C\left(i_t, k_t^{-1}, k_{t-1}^{-1}; \theta, \nu\right) - i_t \quad (2)$$

$$\pi(k_t, \xi_t) = \xi_t k_t^{\alpha} \quad (3)$$

$$C\left(i_t, k_t^{-1}, k_{t-1}^{-1}; \theta, \nu\right) \equiv \frac{\gamma}{2}\left(\frac{i_t}{k_t} - \theta\frac{i_{t-1}}{k_{t-1}} - \nu\right)^2 k_t \quad (4)$$

$$k_{t+1}a_{t+1} = i_t + (1-\delta)k_t \quad (5)$$

$$d_t \geq 0 \quad (6)$$

$$k_0 > 0 \text{ given.} \quad (7)$$

All variables are normalized by the GDP of infrastructure construction (Y_t). Consequently, a_{t+1}, which represents the gross long-term growth rate of value added by the industry, is determined by the following algebraic expression defining the law of motion for capital:

$$\frac{K_{t+1}}{Y_{t+1}}\frac{Y_{t+1}}{Y_t} = \frac{I_t}{Y_t} + (1-\delta)\frac{K_t}{Y_t} \quad (8)$$

where $a_{t+1} \equiv \frac{Y_{t+1}}{Y_t}$; $k_{t+1} \equiv \frac{K_{t+1}}{Y_{t+1}}$; $i_t \equiv \frac{I_t}{Y_t}$. The non-negativity condition $(d_t \geq 0)$ in the system of equations represents a constraint on credit access or external financial frictions faced by the firm. The parameter β denotes the intertemporal discount rate of the representative CEO, who may also be the firm's owner. The operator $\mathbb{E}_t$ reflects the rational expectations of managers, conditional on the information available at time t. Meanwhile, $\pi(\cdot)$ represents the firm's production or technology function. The parameter α represents the proportion of the firm's income allocated to the capital factor (k_t). The variable i_t refers to investment expenditures in civil works and non-residential construction, which are connected to the capital stock through a standard inventory equation and a depreciation rate (δ).

The function $C(\cdot)$ represents the adjustment cost associated with the formation of capital stock, with its parametrization based on Ref. [5]. The parameter γ serves as a measure of the severity of adjustment costs, while θ aims to capture the degree of investment irreversibility by reflecting the persistence of the investment-to-capital ratio. Following Love [5], the rationale for incorporating this term is that it may be optimal for the firm to continue investing a fraction θ of the amount invested in the previous period. Halting or reversing progress on ongoing projects could incur significant costs. In this context, establishing or formalizing long-term investment plans contributes to reducing uncertainty surrounding the firm's expected future returns. Therefore, under conditions of certainty, this parameter is expected to take on values closer to one.

The parameter ν reflects the presence of economies of scale, where costs can be minimized once investment exceeds a certain threshold. Finally, ξ_t represents a variable that measures the total factor productivity (TFP) of the construction firm.

The firm's first-order conditions are derived by applying the Lagrange method to maximize the discounted sum of dividend payments, subject to non-negative dividend payments:

$$max \ \mathcal{L} \equiv \mathbb{E}_t \sum_{s=0}^{+\infty} \beta^s (1+\lambda_{t+s}) d_{t+s} \tag{9}$$

At $s=0$, the target variable is $\{i_t\}$:

$$d_t = \pi(k_t, \xi_t) - C\left(i_t, k_t^{-1}, k_{t-1}^{-1}; \theta, \nu\right) - i_t \tag{10}$$

Therefore, taking into account that $d_t \geq 0$, the Kuhn-Tucker conditions are:

$$\frac{\partial \mathcal{L}}{\partial d_t}\frac{\partial d_t}{\partial i_t} = -(1+\lambda_t)\left(\frac{\partial C}{\partial i_t}+1\right) + \beta(1+\lambda_{t+1})\Phi \leq 0 \tag{11}$$

$$\Phi \equiv \left(\frac{\partial \pi}{\partial k_{t+1}}\frac{\partial k_{t+1}}{\partial i_t} - \frac{\partial C}{\partial i_{t+1}}\frac{\partial i_{t+1}}{\partial k_{t+1}}\frac{\partial k_{t+1}}{\partial i_t} - \frac{\partial i_{t+1}}{\partial k_{t+1}}\frac{\partial k_{t+1}}{\partial i_t}\right) \tag{12}$$

$$d_t \frac{\partial \mathcal{L}}{\partial d_t}\frac{\partial d_t}{\partial i_t} \equiv d_t \frac{\partial \mathcal{L}}{\partial i_t} = 0 \tag{13}$$

$$\frac{\partial \mathcal{L}}{\partial \lambda_t} = \pi(k_t, \xi_t) - C\left(i_t, k_t^{-1}, k_{t-1}^{-1}; \theta, \nu\right) - i_t \geq 0 \tag{14}$$

$$\lambda_t \frac{\partial \mathcal{L}}{\partial \lambda_t} = 0 \tag{15}$$

$$\lambda_t \geq 0 \tag{16}$$

On the other hand, by replacing $\pi(k_t, \xi_t)$ and $C\left(i_t, k_t^{-1}, k_{t-1}^{-1}; \theta, \nu\right)$ with their explicit equivalents, we obtain:

$$\frac{\partial \pi}{\partial k_{t+1}} = \alpha \xi_t k_{t+1}^{\alpha-1} \tag{17}$$

$$\frac{\partial k_{t+1}}{\partial i_t} = \frac{1}{a_t} \tag{18}$$

$$\frac{\partial C}{\partial i_t} = \gamma\left(\frac{i_t}{k_t} - \theta\frac{i_{t-1}}{k_{t-1}} - \nu\right) \tag{19}$$

$$\frac{\partial i_{t+1}}{\partial k_{t+1}} = (1-\delta) \tag{20}$$

$$\frac{\partial C}{\partial i_{t+1}} = \gamma\left(\frac{i_{t+1}}{k_{t+1}} - \theta\frac{i_t}{k_t} - \nu\right) \tag{21}$$

$$\frac{\partial \mathcal{L}}{\partial \lambda_t} = \xi_t k_t^{\alpha} - \frac{\gamma}{2}\left(\frac{k_{t+1}}{k_t}a_t - \theta\frac{k_t}{k_{t-1}}a_{t-1} - (1-\theta)(1-\delta) - \nu\right)^2 k_t - k_{t+1} + (1-\delta)k_t \tag{22}$$

Finally, the optimality condition for the firm's investment decision is derived. The intertemporal investment equilibrium equation differs from that presented by Love [5]. This difference arises because all variables are expressed in terms of sectoral Gross domestic product (GDP), which simplifies the comparison across sectors. This formulation simplifies the calibration of the structural model's parameters. As a result, the first-order optimality conditions include the term a_{t+1} (the long-term growth rate of the infrastructure subsector).

$$d_t\frac{\partial \mathcal{L}}{\partial d_t}\frac{\partial d_t}{\partial i_t} \equiv d_t\frac{\partial \mathcal{L}}{\partial i_t} = 0 \tag{23}$$

$$\gamma\left(\frac{i_t}{k_t} - \theta\frac{i_{t-1}}{k_{t-1}} - \nu\right) + 1 = \beta\frac{(1+\lambda_{t+1})}{(1+\lambda_t)}\Phi \tag{24}$$

$$\Phi = \frac{1}{a_{t+1}}\left[\alpha\xi_t k_{t+1}^{\alpha-1} + \frac{(1+\delta)}{a_{t+1}}\left(\gamma\left(\frac{i_{t+1}}{k_{t+1}} - \theta\frac{i_t}{k_t} - \nu\right) + 1\right)\right] \tag{25}$$

$$i_t = k_{t+1}a_{t+1} - (1-\delta)k_t \tag{26}$$

Finally, based on the first-order condition of the dynamic optimization problem, the steady-state equations are defined:

$$1 + \gamma\left(\frac{i}{k}(1-\theta) - \nu\right) = \beta\frac{(1+\lambda)}{(1+\lambda)}\Phi \tag{27}$$

$$\Phi = \frac{1}{a}\left[\alpha k^{\alpha-1} + \frac{(1-\delta)}{a}\left(1 + \gamma\left(\frac{i}{k}(1-\theta) - \nu\right)\right)\right] \tag{28}$$

$$k = \left[\frac{1}{\alpha}\left(\frac{a\left(1+\gamma\left(\frac{i}{k}(1-\theta)-\nu\right)\right)}{\beta} - \frac{(1-\delta)}{a}\left(1+\gamma\left(\frac{i}{k}(1-\theta)-\nu\right)\right)\right)\right]^{\frac{1}{\alpha-1}} \tag{29}$$

$$i = ka - (1-\delta)k \Rightarrow \frac{i}{k} = \delta + a - 1 \tag{30}$$

3. Parameter calibration

The parameter values in the model are determined through a combination of calibration methods. These methods use statistical data from the National Accounts (CCNN) of the Central Bank of Chile, figures drawn from the economic literature, and original estimates based on Bayesian approximation techniques. This approach was used because not all required time series are available for a long enough historical period. Furthermore, the model incorporates a single source of shock (ξ_t), which helps address the issue of severe stochastic singularity.

The calibration of parameter β takes into account the neutral nominal monetary policy rate, country risk, and an annual inflation target of 3% [12]. The neutral nominal monetary policy rate is from the December 2022 Monetary Policy Report (IPoM), published by the Central Bank of Chile [12]. The report sets the rate at 3.75% annually. Country risk is approximated using Chile's EMBI index, averaging 145 basis points over the 2003–2021 period [13, 14]. Based on this data, the calibrated value for the intertemporal discount factor is:

$$\beta = \frac{1}{1 + 0.0375 + 0.0145 - 0.03} = 0.978 \tag{31}$$

This value is consistent with those commonly found in DSGE macroeconomic models [11]. The parameter α, which represents the contribution of capital to the output of the infrastructure subsector, is calibrated using data from the National Accounts. This value is defined as the ratio between the cost of the remaining capital stock (net of depreciation) and construction GDP. The reference year for the National Accounts is 2018. The remaining capital stock refers to the aggregate stock, excluding residential buildings, machinery, and equipment. The marginal cost of capital (CmgK) is approximated by the interest rate, calculated as $CmgK = 1/\beta$.

We normalize the cost of infrastructure capital by construction GDP because public infrastructure activities are less volatile compared to residential building activities (MACh Reports, CChC[1]). As a result, the construction GDP trend aligns more closely with infrastructure works than with residential construction activities [4, 15]. The parameter α, representing the capital factor payment relative to infrastructure GDP, is approximately 0.526. This indicates that the production process in the infrastructure sector is highly capital-intensive.

The depreciation of the infrastructure capital stock assumes a useful life of 40 years. For tax purposes, the Chilean Internal Revenue Service (SII) defines a useful life of 40 to 80 years for non-residential construction. For engineering works, the useful life ranges from 30 to 75 years. Toledo [16] assumes a uniform useful life of 40 years for both types of construction, drawing on data from the Chilean Internal Revenue Service (SII) and a review of the minimum and maximum useful life spans for non-residential capital stock (35–70 years) and engineering works (330–68 years) across a sample of OECD member countries. Based on this analysis, Toledo [16] estimates an annual depreciation rate (δ) of 2.5%.

This parameter provides a summary measure of the infrastructure subsector, using official data sources from Chile. Although the depreciation rate can vary significantly depending on the type of project—for example, dams or irrigation systems may have a useful life of 150 to 200 years. However, the rate of convergence to the optimal equilibrium with $\delta < 2.5\%$ remains largely unchanged (see Appendix A).

Calibrating the parameter for the adjustment cost to capital formation (γ) is a complex task and is not extensively documented in the Chilean literature. Therefore, the value $\gamma = 4$, estimated by Garcia-Cicco et al. [11], was adopted. This same parameter value is also used by Salas [17] to analyze the macroeconomic impact of income tax reform in Chile. The parameter θ, which reflects the persistence of investment in capital, is derived from Love [5]. This study includes Chile in a panel of 40 countries and empirically demonstrates the causal link between financial markets, economic growth, and firms' investment efficiency.

Furthermore, the long-term gross growth rate of real infrastructure activity (a) is based on the trend gross growth rate of the economy ($\alpha = 1.026$ annually). This

[1] Available from: Centro de Información CChC.

Parameters	Description	Values	Sources
β	Intertemporal discount rate	0.978	Author's calibration based on data from [12–14].
α	Share of infrastructure capital stock	0.526	Author's calibration based on data from BCCh [18]
δ	Annual depreciation rate	1/40	[16]
γ	Adjustment cost for capital formation	4	[11, 17]
θ	Persistence of capital investment	0.215	[5]
a	Gross LP growth rate in the sector	1.026	[12].

Source: Author's analysis.

Table 1.
Calibrated parameters.

approximation stems from economic theories of balanced macroeconomic growth. This trend growth rate is sourced from the December 2022 IPoM[2]. Finally, the parameter ν is defined endogenously under the assumption that, in the steady state (long-term scenario), there are no real frictions in the capital stock of the sector. This leads to the definition $\nu = (\delta + a - 1)(1–\theta)$.

These parameter values enable the quantification of the steady-state ratio between the level of infrastructure capital and the GDP of the same sector. Let K_{inf} denote the capital stock in infrastructure works and Y_{inf} the output produced by firms in this sector.

Thus, $k = \frac{K_{inf}}{Y_{inf}} = 63.8$, indicating that the capital stock in civil works and non-residential buildings is 63.8 times the value added by this construction subsector. This result is consistent with data from the National Accounts (CCNN). Let Y and Y_c represent total GDP and construction GDP, respectively. Based on National Accounts statistics (with 2018 as the reference year), the following relationships are observed:

$\frac{K_{inf}}{Y} = 1.68$; $\frac{Y_c}{Y} = 6.5\%$ (construction GDP as a share of total GDP). The contribution of engineering and non-residential building projects to the total construction sector averages 40.5%. Accordingly, the following observations can be made:

$$\frac{K_{inf}}{Y_{inf}} = \frac{K_{inf}}{Y}\left(\frac{1}{\frac{Y_c}{Y}}\right)\left(\frac{1}{\frac{Y_{inf}}{Y_c}}\right) = 1.68\left(\frac{1}{6.5\%}\right)\left(\frac{1}{40.5\%}\right) = 63.8 \tag{32}$$

Table 1 summarizes the parameter values of the structural model:

3.1 Identification strategy

To statistically validate the calibrated values of β and α, which were not derived from formal research sources, we employed Bayesian estimation techniques utilizing Monte Carlo simulations with Markov chains (MCMC). This methodology allows us to infer parameter distributions by integrating observed system variable data with prior

[2] Available at the Central Bank of Chile.

information [19]. The prior information may originate from the calibration process or alternative microeconomic sources [20]. Combining data with prior assumptions not only facilitates the convergence of the Metropolis-Hastings optimization algorithm in MATLAB but is particularly advantageous when working with relatively short time series, a characteristic often observed in sectoral national account statistics across Latin America [21].

The parameter estimation uses the annual series of gross fixed capital formation, excluding housing services and mining, published by the Central Bank of Chile for the period 2008–2021. This series represents real investment in infrastructure projects (I_t), specifically in civil works and non-residential construction.

Figure 1 compares the trajectory of this series with its trend, estimated using the Hodrick-Prescott (HP) filter. **Figure 2** shows the percentage gap in infrastructure investment, calculated as the logarithmic difference between the variable It and its HP trend.

Several key observations emerge. In 2009, infrastructure investment fell approximately 11% below its trend, taking nearly 2 years to recover, largely due to the global financial crisis and the February 27, 2010, earthquake. From 2015 onward, the cycle reveals a declining trend, reflecting structural weaknesses in investment dynamics despite some volatility. The COVID-19 pandemic in 2020 caused widespread economic disruption; however, policy interventions in Chile mitigated the decline compared to the 2009 recession and accelerated recovery in 2021. Nevertheless, these stimulus measures introduced significant inflationary pressures in the years that followed. Consequently, the reactivation measures implemented in 2020–2021 have normalized, bringing to light persistent structural issues suppressing aggregate investment in the economy. Infrastructure investment, in particular, remains critical, accounting for nearly two-thirds of total investment.

Figure 3 depicts the variable $i_t \equiv I_t/Y_t$ from the model, where Y_t represents sectoral GDP measured in real terms at constant 2018 prices. This time series is utilized for the Bayesian parameter estimation, as it aligns with the definitions of the model variables outlined in Section 2.

Table 2 presents the estimated mean values of the parameters α and β, along with their 95% confidence intervals and the associated prior distributions. Additionally,

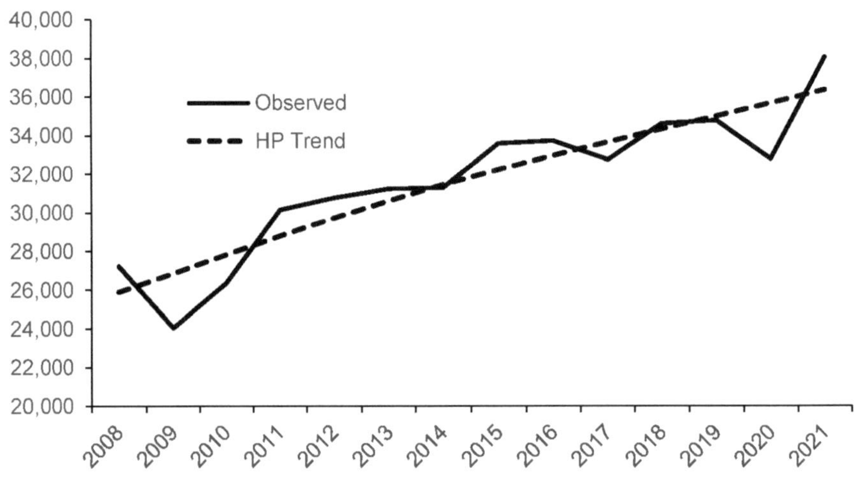

Figure 1.
Infrastructure investment (millions of US$). Source: Author's analysis.

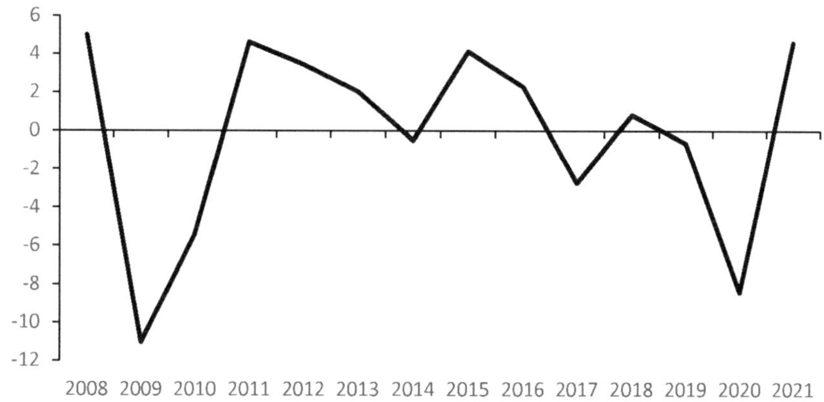

Figure 2.
Infrastructure investment gap (percentage). Source: Author's analysis.

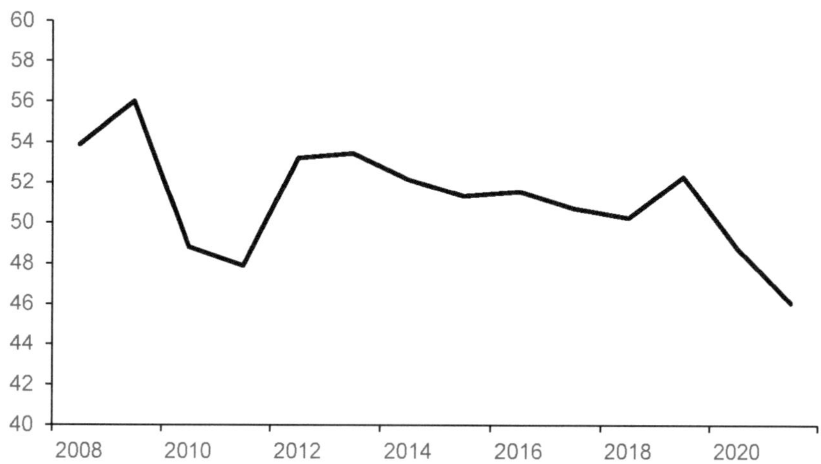

Figure 3.
Evolution of i_t: Infrastructure investment normalized by sectoral GDP (in percentage). Source: Author's analysis.

Parameter	Mean (Prior)	Mean (Posterior)	95% CI	Distribution
β	0.970	0.960	0.940–0.980	*Normal*
α	0.600	0.565	0.477–0.644	*Beta*
ρ	0.700	0.924	0.876–0.975	*Beta*
σ	0.500	0.349	0.209–0.488	*Inverse Gamma*

Source: Author's analysis.

Table 2.
Estimated parameters.

two supplementary parameters (ρ and σ) were estimated, characterizing the stochastic equation for the productivity shock (ξ_t) in the production or technology function of the representative firm. The stochastic equation, incorporated into the system to estimate the parameters of interest, is detailed below.

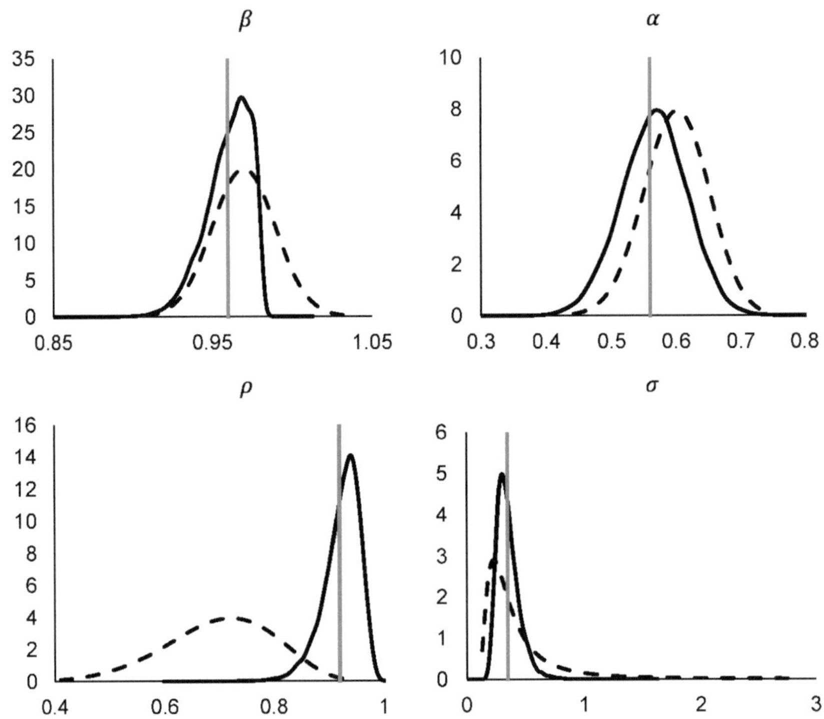

Figure 4.
Distributions of estimated parameters. Note: The segmented curve represents the prior distribution, while the solid curve represents the posterior distribution. The vertical gray line indicates the mean of the posterior distribution. Source: Author's original work.

$$\xi_t = \rho\xi_{t-1} + \varepsilon_t, \varepsilon_t \sim^{idd} \left(0, \sigma^2\right) \tag{33}$$

The results in the table indicate that the calibrated values of the parameters α and β presented in **Table 1** lie within the 95% confidence intervals and are closely aligned with their respective posterior mean values. Moreover, the estimated values are statistically significant, as their confidence intervals do not include zero.

Additionally, **Figure 4** illustrates that the available observations on infrastructure investment provide valuable information for parameter identification, as the estimated values are derived from a combination of the prior distribution and the posterior distribution informed by the data.

4. Simulations

Using Bellman equations (dynamic programming) to solve the optimization problem yields the following results:

i. The value function of the representative firm in the infrastructure subsector, which represents the maximum expected payoff for each state of the capital stock, expressed as multiples of real infrastructure GDP.

ii. The policy function, illustrating the transition of the infrastructure capital stock toward its long-term equilibrium level.

iii. The effect of adjustment costs on capital formation and their influence on the convergence rate to equilibrium.

Formally, the Bellman equation for this model is represented as follows:

$$V_t(k_t, \xi_t) = d_t + \max_{\{i_{t+s}\}_{s=1}^{+\infty}} \mathbb{E}_t \sum_{s=1}^{+\infty} \beta^s d_{t+s} \tag{34}$$

$$V_t(k_t, \xi_t) = d_t + \beta\left(\max_{\{i_{t+s}\}_{s=0}^{+\infty}} \mathbb{E}_t \sum_{s=0}^{+\infty} \beta^{s+1} d_{t+s+1} \right) \tag{35}$$

$$V_t(K_t, \xi_t) = d_t + \beta V_{t+1}(k_{t+1}, \xi_{t+1}) \tag{36}$$

$$V(K, \xi) = d + \beta V'(k', \xi') \tag{37}$$

Subject to:

$$d = \xi k^a - \frac{\gamma}{2}\left(\frac{i}{k} - \theta\frac{i}{k_{-1}} - \nu\right)^2 k - i \tag{38}$$

$$k'a = i + (1-\delta)k \tag{39}$$

$$d \geq 0 \tag{40}$$

The dynamic optimization method to solve this problem was implemented in MATLAB R2022a. **Figure 5** shows that the firm's value function has the expected concave shape, as described in the economic literature. This function represents the relationship between the maximum expected value of the representative firm and the capital stock resulting from its investments. Its concavity reflects the risk aversion inherent to construction firms, which is consistent with the irreversible nature of such investments.

Specifically, there is no secondary market where partially completed construction projects can be sold at market value. In other words, once a project begins, there is no opportunity to reverse the decision without incurring sunk costs [22].

The policy function depicted in **Figure 6** illustrates the transition of capital toward its equilibrium or long-term level, expressed as multiples of real infrastructure GDP. The figure demonstrates that capital gradually converges to its optimal equilibrium level, influenced by frictions in the implementation of infrastructure projects. These frictions, modeled as adjustment costs associated with capital formation, align with the established empirical observations.

For instance, research studies [1, 2, 23] underscore that in Chile, the lack of explicit mechanisms and criteria for selecting construction modalities (e.g., concessions vs. traditional public works) substantially delays the progression of infrastructure projects. Reflecting these adjustment costs in the model, the steady-state prediction indicates that the capital stock level is approximately 64 times the infrastructure GDP.

In 2021, for example, the capital stock for civil works and non-residential buildings was about 62 times the GDP of the same subsector. Since this value falls short of the equilibrium level estimated by the model, **Figure 6** displays an upward-sloping convergence path for the capital stock. This trajectory suggests that the capital stock, normalized by infrastructure GDP, is progressing toward its long-term equilibrium level.

This section analyzes the simulations generated using a range of arbitrary values for the adjustment cost parameter (γ) in the capital formation function. Notably, deviations

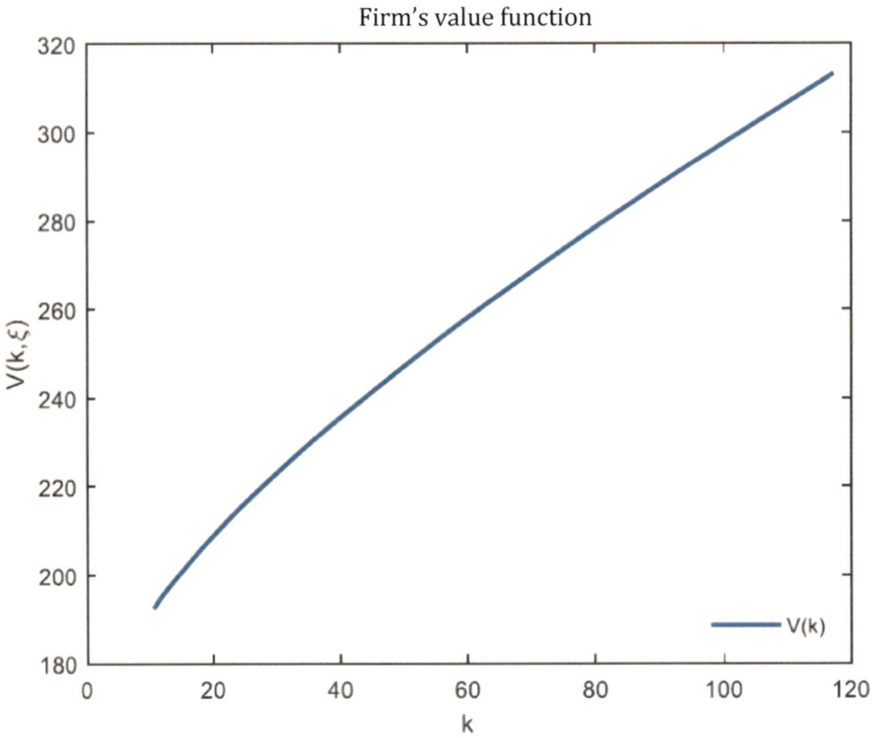

Figure 5.
Relationship between the firm's maximum expected value ($V_t(k_t, \xi_t)$) and the capital stock derived from its investments (k_t). Source: Author's analysis.

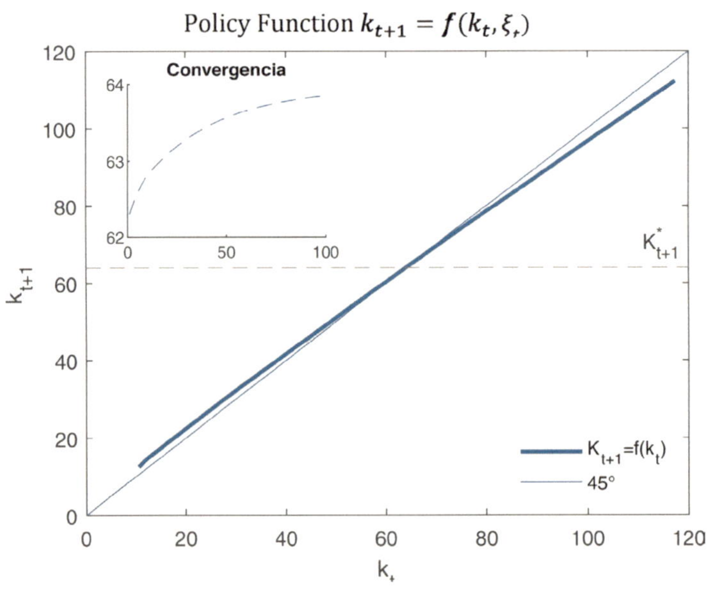

Figure 6.
Policy function. Source: Author's analysis.

above the baseline scenario parameter ($\gamma = 4$) indicate a greater degree of rigidity or inefficiency in Chile's infrastructure investment process. Conversely, when $\gamma \rightarrow 0$, the model assumes a perfectly efficient market with no frictions in the investment process.

Accordingly, factors such as the regulatory framework—intended to ensure legal security in contracts between public and private entities—the process for obtaining permits and licenses for project initiation, and the effective management of infrastructure-related risks in agreements between contractors and public-sector entities are incorporated into the adjustment cost variable (γ) in the model. This interpretation of γ is supported by productivity studies in the construction sector, such as those conducted in research studies [1, 2]. For instance, the CNEP study found that between 60 and 70% of the frictions or inefficiencies in civil works investments are attributable to state bureaucracy. Consequently, state actions are crucial for sustaining infrastructure investments.

Figures 7 and **8** illustrate that a reduction in the parameter (γ), associated with the adjustment cost of investment, relative to the baseline scenario, accelerates the convergence of the capital stock toward its long-term equilibrium. Conversely, an increase in γ reduces the rate of convergence. Specifically, the accumulation of infrastructure stock is delayed when the processes required to execute projects become overly burdensome. Consequently, government measures that decrease the adjustment cost of infrastructure formation by 25–50% compared to the baseline scenario ($\gamma = 4$) can reduce execution times by 15–30%. This translates to a reduction in the average completion time for major infrastructure projects from 12 years (current baseline) to a range of 6 to 10 years. A policy recommendation to facilitate the advancement of stalled projects and promote new investment initiatives is the establishment of an independent infrastructure agency. Such an agency would operate beyond political cycles, focusing on the medium- and long-term planning and prioritization of projects while also defining formal criteria for contracting modalities between public authorities and contractors.

Figures 9 and **10** focus on the parameter θ, which represents the persistence of investment. For Chile, this parameter is calibrated at 0.215, reflecting low investment

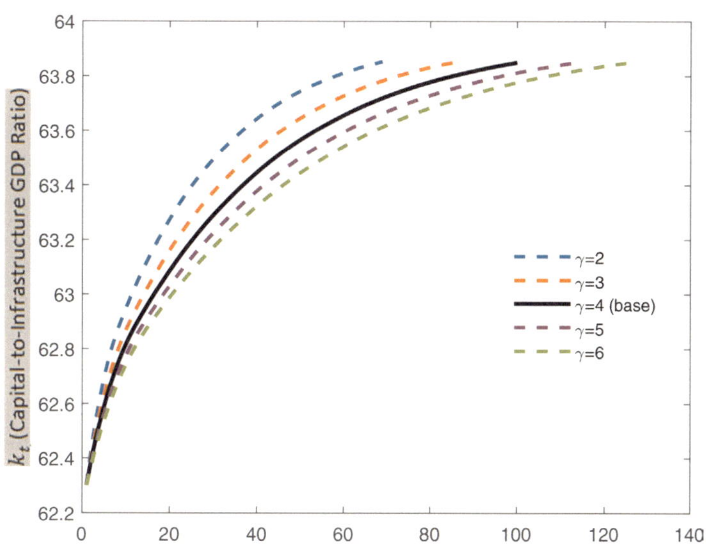

Figure 7.
Impact of adjustment costs on infrastructure capital formation. Source: Author's analysis.

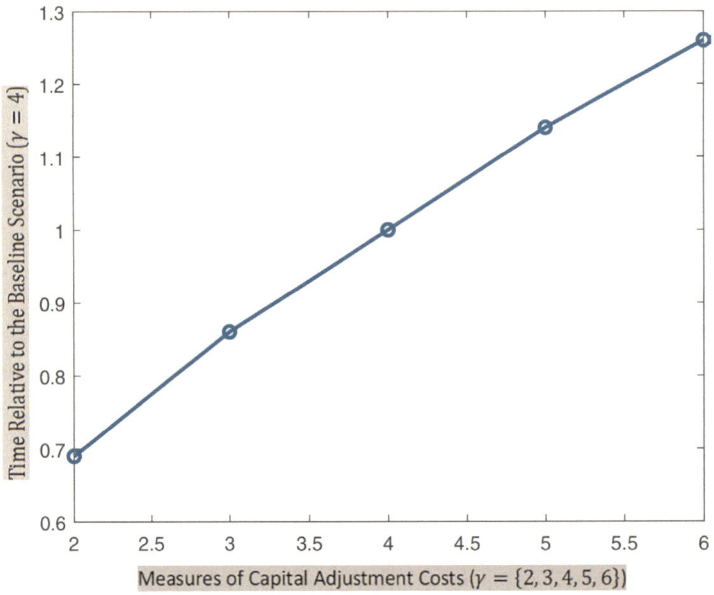

Figure 8.
Convergence time of capital stock to long-term equilibrium under different adjustment cost measures. Source: Author's analysis.

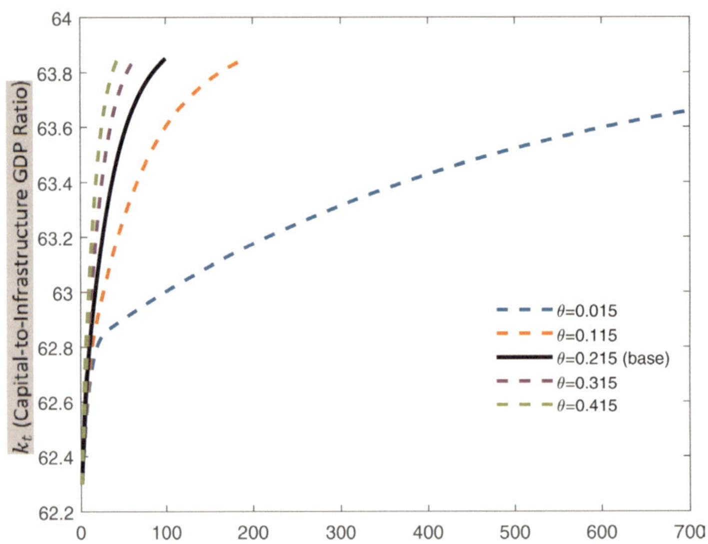

Figure 9.
The impact of sustained investment. Source: Author's analysis.

continuity in infrastructure projects compared to a panel of over 40 countries analyzed by Love [5]. Implementing medium- and long-term infrastructure investment policies could enable firms to optimally allocate a portion of current investments toward future projects. For example, firms could dedicate additional resources to workforce training and the adoption of new technologies, fostering the industrialization of construction processes. In this scenario, productivity improvements in the

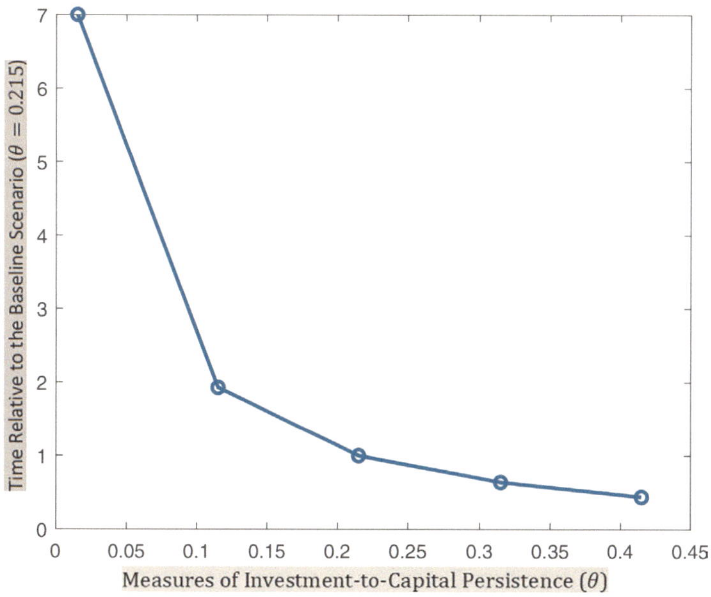

Figure 10.
Relative convergence time in relation to investment persistence. Source: Author's analysis.

sector would be driven by state actions that provide greater certainty regarding the continuity of medium- and long-term investment plans or projects.

Ensuring the continuity of investment mitigates the adverse effects of bureaucracy or inefficiencies within the broader state apparatus. The model predicts that as investment becomes more persistent ($\theta \to 1$), project execution times decrease, resulting in faster convergence of capital to its optimal equilibrium level—paralleling the sensitivity observed using γ (**Figures** 7 and **8**).

For instance, increasing investment continuity by raising the persistence parameter from 0.22 to a range of 0.27 to 0.32—equivalent to a rise from 25–50%—could reduce the time required to accumulate the capital stock by 19–36%. While this scenario has a comparable impact on construction timelines to adjustments in the cost of investment, the effort and cost of establishing future-oriented investment plans that ensure continuity, irrespective of economic or political cycles, are significantly lower than the resources required to enhance the productivity of the state as a whole. Achieving the latter would entail substantial regulatory reforms and complex political agreements.

In this context, the model provides formal support for the importance of adopting an infrastructure investment policy that is independent of economic cycles. Such policies can be implemented in the short term and yield outcomes comparable to those derived from improving overall state efficiency (sensitivity of γ). Accordingly, a strategic starting point would be the institutionalization of comprehensive medium- and long-term investment plans, such as the establishment of an infrastructure agency, while concurrently advancing efforts to modernize state functions. Establishing structural investment plans necessitates concrete actions with a demonstrable short-term positive impact on business confidence, in contrast to the formidable challenge of improving state productivity as a whole, which requires highly complex legislative reforms.

4.1 International evidence on institutional planning and investment impact

A substantial body of international evidence supports the argument that robust institutional mechanisms for long-term infrastructure planning and prioritization lead to significant increases in both public and private investments, while improving execution efficiency. **Table 3** summarizes key examples from countries and regions that have adopted such frameworks, highlighting their observed quantitative impacts, institutional tools, and critical success factors.

4.2 Policy recommendations

To establish a robust foundation for long-term planning and foster an environment conducive to its effective implementation, it is recommended that both the State and regulatory bodies consider the adoption of complementary measures. These additional actions can play a significant role in promoting sustained infrastructure investment, thereby supporting sustainable development. Key recommendations include:

4.2.1 Strategic long-term planning and future demand assessment

This involves identifying the infrastructure and investment requirements necessary to meet future demand. Developing a comprehensive infrastructure plan that addresses both current and future needs, sets clear objectives, and prioritizes development areas is essential. Effective planning enables the identification of opportunities and the proactive preparation of infrastructure projects.

Country/ region	Institutional instrument	Year Created	Estimated Quantitative Impact	Key Success Factors	Source
United Kingdom	*National Infrastructure Commission* (NIC) —Five-Year Assessment	2015	+0.6 percentage points of GDP in annual infrastructure investment (2021–2025)	Technical independence, 30-year horizon, prioritized portfolio	[24]
Australia	*Infrastructure Australia*—15-Year Plan + Priority List	2008	−25% project cancelations; +0.4 to 0.7% of GDP	Legal framework, mandatory cost–benefit analysis, public transparency	[25]
Canada	*Canada Infrastructure Bank* (CIB)	2017	CAD $1 yields CAD $2.5–3.5; +0.5–0.6% of GDP annually	Co-financing, strategic guidance, operational autonomy	[26]
European Union	*Connecting Europe Facility* + Ten-Year Network Plans	Since 2015	+0.3% of GDP annually (2015–2021)	Multiannual funds, regional governance, climate-aligned goals	[27]
Meta-studies	—	—	+0.5 to 2 percentage points of GDP over the long run	Strong institutions reduce risk and improve project execution	[28, 29]

Source: Author's analysis.

Table 3.
International evidence on institutional planning and investment impact.

4.2.2 Establishing a clear and stable regulatory framework

Providing a regulatory environment that is clear, stable, and predictable is critical for ensuring legal certainty for investors. This includes enacting laws, regulations, and standards that clearly define technical, operational, environmental, and safety requirements for infrastructure construction and operation. Simplifying administrative procedures, clarifying processes for obtaining permits and licenses, and ensuring transparency in environmental and labor regulations are crucial. Additionally, introducing tax stability agreements, particularly for strategic sectors or climate change initiatives, and strengthening the legal framework underpinning project regulations and permits will create greater certainty and facilitate infrastructure investment.

4.2.3 Implementing public investment programs with a crowding-in effect

Public investment programs in infrastructure can mobilize public funds to finance high-priority projects. Such investments have the potential to enhance regional capacity to support productive activities, thereby accelerating private sector project pipelines. Strengthening the financial autonomy of subnational governments is also essential to achieving this goal.

4.2.4 Promoting investments in productive infrastructure

Related to the previous point, public investment programs should focus on improving transportation, energy, and water infrastructure in productive areas. Emphasizing connectivity and resource access projects can reduce logistical costs, enhance economic viability, and increase the attractiveness of projects to investors.

These recommendations align with the 2017 OECD report, [30]. By implementing consistent and well-coordinated actions, governments can facilitate a steady flow of infrastructure projects, improving the quality, efficiency, and sustainability of their portfolios. This approach addresses inefficiencies in capital formation caused by bureaucratic hurdles, ultimately enabling investment levels that foster sustainable economic growth, improved quality of life, and the resolution of pressing societal needs [31].

4.3 Further discussion: Institutional, comparative, and stakeholder dimensions

Although the model is calibrated specifically for the Chilean economy, its findings can be relevant to other countries facing infrastructure bottlenecks caused by institutional or planning inefficiencies. A comparative extension with peer economies—such as Peru, Colombia, or Mexico—could enrich the analysis by contrasting the effectiveness of different institutional arrangements for long-term infrastructure investment. Furthermore, while Bayesian estimation was employed to enhance credibility, conducting sensitivity analyses on key assumptions—such as adjustment costs or investment persistence—would improve the robustness of the results across varying economic conditions. These represent important avenues for future research.

The interpretation of adjustment costs as a catch-all proxy for various institutional and contractual frictions was a deliberate modeling choice for the sake of parsimony. However, we acknowledge that this simplification may obscure significant differences across friction types (e.g., environmental, fiscal, regulatory). Disaggregating these factors could improve future versions of the model. Similarly, the proposal to create an autonomous Infrastructure Agency draws on global best practices, but it still requires a more thorough analysis of political feasibility, current institutional capacities, and the likely resistance of key stakeholders (such as ministries, subnational governments, and contractors). Incorporating perspectives from local communities, labor unions, and environmental actors would also broaden the approach and enhance the legitimacy and sustainability of proposed reforms. Finally, outlining a phased implementation roadmap for the agency, including specific milestones, would improve the real-world applicability of the policy recommendations[3].

5. Conclusions

Ensuring the consistent development of investment plans for infrastructure projects is a key measure to mitigate inefficiencies in capital formation caused by excessive bureaucracy. Enhancing state productivity and reducing adjustment costs by 25–50% compared to the baseline scenario ($\gamma = 4$) could decrease the execution time of civil works and non-residential buildings by 15–30%. This would reduce the average completion time for major infrastructure projects from 12 years to a range of 6 to 10 years.

Moreover, maintaining investment continuity offsets the negative impacts of bureaucratic inefficiencies within the State system. As investment becomes more persistent ($\theta \to 1$), the model predicts significant reductions in project execution times. In effect, ensuring the continuity of investments accelerates capital convergence to its optimal equilibrium level, yielding benefits comparable to a substantial improvement in State productivity.

This finding underscores the critical importance of implementing long-term planning policies for infrastructure investment. It also highlights that establishing a structural or long-term investment planning mechanism requires more concrete and actionable steps than the broader challenge of modernizing the State as a whole.

Future research could explore the relative impact of two policy measures on investment continuity:

1. Accelerating the approval process for environmental permits.

2. Establishing regulations that enhance the legal certainty of approved permits.

Which measure would have a greater impact: expediting permit approvals or ensuring the legal certainty of granted permits?

Appendix 1: Convergence with variations in the useful life of capital

Figure 11 shows minimal variability in the convergence times of the capital stock to its long-term equilibrium when the capital depreciation rate is below 2.5% ($\delta \leq 1/40$).

[3] Appendix B proposes a preliminary baseline to outline a gradual implementation strategy for the proposed agency, with phases and institutional milestones, which would facilitate its feasibility in the Chilean context.

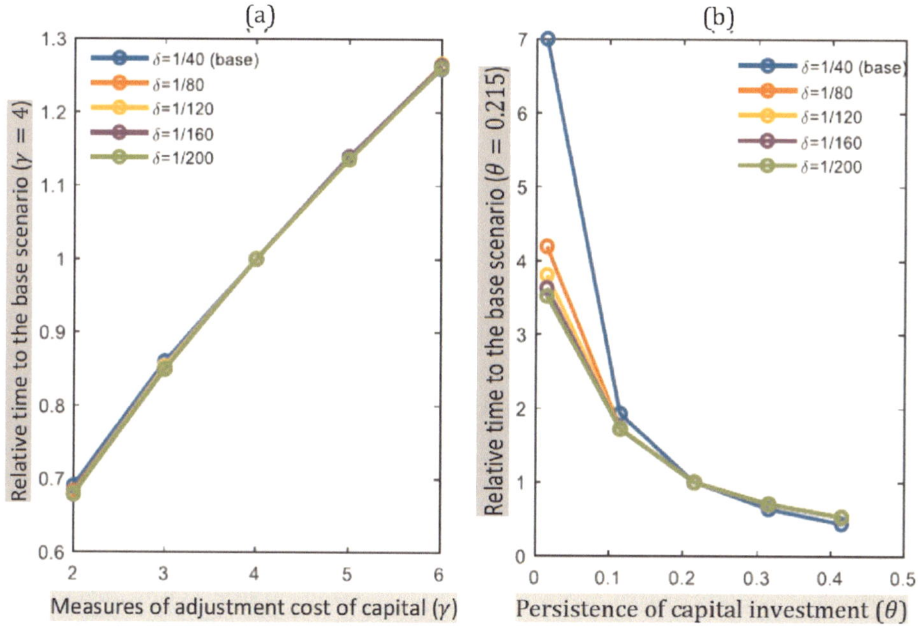

Figure 11.
Convergence with variations in the useful life of capital. (a) Time to convergence of the capital stock relative to its adjustment cost. (b) Time to convergence of the capital stock relative to the persistence of investment. Source: Author's analysis.

Appendix 2: National System for Long-Term Infrastructure Planning and Investment: A preliminary public policy proposal for Chile

General objective

To establish an institutionalized, technical, and transparent system to guide the planning, prioritization, financing, and evaluation of public and public-private infrastructure projects with a State vision, promoting spending efficiency and accelerating the execution of social and strategic projects.

Key components of the proposal

1. Creation of a National Infrastructure Agency (ANI): Inspired by Infrastructure Australia and the UK's National Infrastructure and Construction Pipeline (NISTA)

Attributions:

- Develop a 30-year National Infrastructure Plan, updated every 5 years.
- Annually publish a Priority Infrastructure Project Portfolio.
- Apply standardized methodologies for social, environmental, and territorial evaluation.

- Coordinate with ministries, regional governments, and municipalities.
- Propose financing schemes (public works, PPPs, social leasing).
- Report to Congress and the public.

Legal nature:

- Autonomous agency, created by law, under the Ministry of Finance or the Presidency.
- Collegiate governance with technical, regional, and sectoral representation.

2.Legal Framework: Long-Term Infrastructure Law

Key contents:

- Establishes ANI's role and mandate.
- Requires that any project over UF 500,000 (≈ USD 17 million) be evaluated using ANI methodology. Projects between UF 100,000 and 500,000 may be evaluated by sectoral units under ANI oversight through simplified protocols.
- Defines prioritization criteria: social profitability, territorial impact, etc.
- Establishes an infrastructure fiscal rule for multi-year funding.

3.National Bank of "Shovel-Ready" Social Projects

Function:

- Maintain a public and dynamic list of high social impact projects (hospitals, rural drinking water systems, schools) with technical designs and permits ready.

Instruments:

- Standard manuals, bidding document templates, framework agreements.

4.Strategic Social Infrastructure Fund: Multi-year budget fund inspired by the EU's Connecting Europe Facility (CEF).

- Finances high social impact projects prioritized by ANI.
- Access is conditioned on ANI standards.
- Enables countercyclical stimulus.

Implementation phases

- Stage I: Diagnosis and Design (6 months): Draft law and institutional design.

- Stage II: Legal Processing (12 months): Approval of Infrastructure Law.
- Stage III: ANI Setup and System Launch (12–18 months): First National Infrastructure Plan published.
- Stage IV: Full Operation (From Year 3): Prioritization of national and regional public portfolio.

Conclusion

Providing Chile with a modern, technical, and stable institutional framework for infrastructure is key to closing social gaps, stimulating investment, and ensuring efficient use of public resources over the next 30 years.

Author details

Byron J. Idrovo-Aguirre[1,2]*, Héctor Acuña[1] and Pablo Easton[1]

1 Department of Research and Public Policy, Chilean Chamber of Construction (CChC), Santiago, Chile

2 Universidad Adolfo Ibáñez, Santiago, Chile

*Address all correspondence to: bidrovo@cchc.cl

References

[1] CNEP. Productividad en el Sector de la Construcción. Chile: National Commission for Evaluation and Productivity, CNEP; 2020

[2] CChC-Matrix. Impulsar la Productividad de la Industria de la Construcción en Chile a Estándares Mundiales. Chilean Chamber of Construction and Matrix Consulting; 2020. Available from: https://construye2025.cl/wp-content/uploads/2022/07/Resumen-Ejecutivo-Estudio-de-Productividad-Construccio%CC%81n-2020.pdf

[3] Bentancor A. Evolución de la Productividad del Sector Construcción entre 1986–2007. Working Paper No. 54, Chilean Construction Chamber; 2009. Available from: https://drive.google.com/drive/folders/1QIeb5ngMAraDl-GGaZYuGWJKecno7Ji0?usp=drive_link

[4] Idrovo B. Inversión en Vivienda e Infraestructura y sus Principales Determinantes Macroeconómicos. Working Paper No. 88, Chilean Construction Chamber; 2018. Available from: https://drive.google.com/drive/folders/1QIeb5ngMAraDl-GGaZYuGWJKecno7Ji0?usp=drive_link

[5] Love I. Financial development and financing constraints: International evidence from the structural investment model. The Review of Financial Studies. 2003;**16**(3):765-791

[6] Gilchrist S, Himmelberg C. Investment: Fundamentals and finance. NBER Macroeconomics Annual. 1998;**13**:223-262

[7] Whited TM. Debt, liquidity constraints, and corporate investment: Evidence from panel data. The Journal of Finance. 1992;**47**(4):1425-1460

[8] Hubbard RG, Kashyap AK. Internal net worth and the investment process: An application to US agriculture. Journal of Political Economy. 1992;**100**(3):506-534

[9] Hubbard RG, Kashyap AK, Whited T. Internal finance and firm investment. Journal of Money, Credit, and Banking. 1995;**27**(3):683-701

[10] Calomiris C, Hubbard G. Internal finance and firm-level investment: Evidence from the undistributed profit tax of 1936-1937. Journal of Business. 1995;**68**:443-482

[11] Garcia-Cicco J, Justel S, Kirchner M. Financial Frictions and the Transmission of Foreign Shocks in Chile. Working Paper No. 722, Central Bank of Chile; 2014. Available from: https://www.bcentral.cl/en/detail-news-and-publications/-/asset_publisher/Exzd7l9NC3Y6/content/documento-de-trabajo-n-722

[12] Central Bank of Chile. Monetary Policy Report (IPoM). Central Bank of Chile; 2022. Available from: https://www.bcentral.cl/en/web/banco-central/content/-/detalle/informe-de-politica-monetaria-diciembre-2022

[13] JPMorgan Chase. Available from: https://www.jpmorgan.com/CL/en/about-us

[14] Central Bank of Chile. Emerging Markets Bond Index (EMBI) Data. Available from: https://si3.bcentral.cl/siete/ES/Siete/Cuadro/CAP_EI/MN_EI11/EI_EMBI?idSerie=F019.SPS.PBP.90.D

[15] Idrovo B, Contreras J. Bayesian monthly index for building activity based on mixed frequencies: The case of Chile. Journal of Economic Studies. 2022;

49(3):541-557. DOI: 10.1108/JES-01-2021-0022

[16] Toledo P. Stock de Capital de la Economía Chilena y su Distribución Sectorial. Working Paper No. 233, Central Bank of Chile; 2003. Available from: https://www.bcentral.cl/documents/33528/133326/DTBC_233.pdf

[17] Salas S. El Impacto Macroeconómico de la Desintegración Tributaria Total. In: Documento de Trabajo N 15, Economic Context Observatory. Diego Portales University; 2022. Available from: https://ocec.udp.cl/proyecto/documento-de-trabajo-no15-el-impacto-macroeconomico-de-la-desintegracion-tributaria-total/

[18] Central Bank of Chile. Capital Stock Data. Available from: https://si3.bcentral.cl/Siete/ES/Siete/Cuadro/CAP_CCNN/MN_CCNN76/CCNN2018_S2_P1_DUPLICADO/638682982910347075

[19] Del Negro M, Schorfheide F. Forming Priors for DSGE Models (and How it Affects the Assessment of Nominal Rigidities). NBER Working Paper w13741; 2008. Available from: https://www.sciencedirect.com/science/article/abs/pii/S0304393208001396

[20] Smets F, Wouters R. An estimated dynamic stochastic general equilibrium model of the euro area. Journal of the European Economic Association. 2003;**1**:1123-1175

[21] Morandé F, Tejada M. Shocks de Oferta Persistentes: ¿Un dolor de cabeza para los bancos centrales? Vol. 12, N 3. Economía Chilena, Central Bank of Chile; 2009. Available from: https://repositoriodigital.bcentral.cl/xmlui/handle/20.500.12580/3521

[22] De Gregorio J. Crisis de Confianza Global: El Valor de Esperar y las Fallas de Coordinación Una Vez Más. Documento de Política Económica N 31, Central Bank of Chile; 2009. Available from: https://www.bcentral.cl/documents/33528/133214/bcch_presentac_163697_es.pdf/ef352574-0b4c-6aa0-e86c-8fcd01322784?t=1698076624433

[23] Acuña H, Easton P. Más allá de Asociaciones Público Privadas (APP): Aspectos claves en la atracción de financiamiento privado para el desarrollo de la infraestructura pública. Working Paper No. 95, Chilean Construction Chamber; 2019. Available from: https://drive.google.com/drive/folders/1QIeb5ngMAraDl-GGaZYuGWJKecno7Ji0?usp=drive_link

[24] HM Treasury. National Infrastructure Strategy. Government of the United Kingdom; 2020. Available from: https://www.gov.uk/government/publications/national-infrastructure-strategy

[25] Productivity Commission. Public Infrastructure: Inquiry Report No. 71. Canberra: Australian Government; 2014. Available from: https://www.pc.gov.au/inquiries/completed/infrastructure/report/infrastructure.pdf

[26] Oxford Economics. Canada Infrastructure Bank: Economic Impact Analysis. Report Prepared for Canada Infrastructure Bank; 2020. Available from: https://cib-bic.ca/

[27] European Commission. Connecting Europe Facility – Annual Report 2022. Directorate-General for Mobility and Transport; 2022. Available from: https://transport.ec.europa.eu/transport-themes/infrastructure-and-investment/connecting-europe-facility_en

[28] Calderón C, Servén L. Infrastructure and economic development in sub-Saharan Africa. Journal of African

Economies. 2010;**19**(suppl_1):i13-i87. DOI: 10.1093/jae/ejp022

[29] OECD. Long-Term Strategic Vision for Infrastructure: OECD Framework. OECD Publishing; 2020. Available from: https://infrastructure-toolkit.oecd.org/governance/strategic-vision/

[30] OECD. Gaps and Governance Standards of Public Infrastructure in Chile: Infrastructure Governance Review. Paris: OECD Publishing; 2017. DOI: 10.1787/9789264278875-en

[31] OECD. Developing Robust Project Pipelines for Low-Carbon Infrastructure, Green Finance and Investment. Paris: OECD Publishing; 2018. DOI: 10.1787/9789264307827-en